COLORADO

T0153044

Rocks & Minerals

A Field Guide to the Centennial State

Dan R. Lynch & Bob Lynch

Adventure Publications
Cambridge, Minnesota

Dedication

To Nancy Lynch, wife of Bob and mother of Dan, for her love and continued support of our book projects.

And to Julie Kirsch, Dan's wife, for her love and patience.

Acknowledgments

Thanks to George Robinson, Ph.D, Jim and Irene Witmer, Jennifer Klava, Brian Costigan and Larry Costigan for providing information and specimens.

Photography by Dan R. Lynch

Cover and book design by Jonathan Norberg

Edited by Brett Ortler

15 14 13 12 11 10 9 8 7

Colorado Rocks & Minerals
Copyright © 2010 by Dan R. Lynch and Bob Lynch
Published by Adventure Publications
An imprint of AdventureKEEN
310 Garfield Street South
Cambridge, Minnesota 55008
(800) 678-7006
www.adventurepublications.net
All rights reserved
Printed in China
ISBN 978-1-59193-238-3 (pbk.)

Table of Contents

Introduction

Colorado is one of the most mineralogically diverse states in the country and has long been the site of exciting and important rock and mineral discoveries. The state owes much of its geological interest and wealth of unique minerals to the Colorado Rocky Mountains, most of which formed when enormous formations of granite forced their way upward through the earth's surface. The mountains divide the state down its center; to the east are flat, low-lying plains created by an ancient sea, and to the west are the high-elevation plateaus, formed in part by volcanic activity. A rich history in mining and exploration has brought Colorado's riches to light, including world-class mineral specimens, billions of dollars' worth of gold, fossils new to science, and even oil. As you can imagine, Colorado is a rock hound's dream destination, made even more special by having the Rockies as a backdrop.

Private and Protected Places

It is your responsibility to know where you can and cannot collect rocks and minerals in Colorado. Colorado has dozens of nationally and state-protected forests, parks and monuments, all of which are areas where it is illegal to collect anything. Large fines await those caught collecting in protected areas. You are encouraged to obey the law and leave the natural spaces wild and untouched for generations to come. While protected lands are often marked on maps, private land is not. Needless to say, if you're caught collecting on privately owned land without permission, the penalty may be worse than just a fine.

Dangerous Places

Along with Colorado's rich mining history come hundreds of mines, both new and old. It is unlikely that you will ever come across an unrestricted mine entrance, but if you do, you must never enter! Age and disrepair may have made many of these mines unstable and prone to collapse. In addition, dangerous gasses can build up in old mines, and suffocation can be the result for foolish rock hounds. Quarries, gravel pits, and mine dumps (piles of waste rock left over at mines) also present their own dangers. Steep piles of loosely compacted rock or gravel can easily start an avalanche if you carelessly climb upon them.

Colorado Weather

In the higher elevation areas of Colorado, the changes in weather can be drastic. A warm, sunny day can abruptly turn into a windy snowstorm with little warning, and you wouldn't want to be trapped on a lonely mountainside when it happens. Weather forecasts for mountainous areas will help, but being prepared by keeping extra clothing and survival equipment in your vehicle is your best defense.

Elevation

Denver, Colorado, may be famous as the "Mile High City," but it is far from being the highest elevation area in the state. If you're visiting from a low elevation area and are planning on hunting rocks in Colorado's mountainous regions, the height may take you by surprise, causing shortness of breath and dizziness when exerting yourself. The air is thin, and if you're not accustomed to it, you can tire very easily. And in cities like Leadville, the highest incorporated city in the US at over 10,000 feet, you'd do well to be careful, work slowly, and take plenty of breaks.

Glossary Note

Books about geology, rocks and minerals can be very technical. To help make this book intuitive for the layperson but useful for the expert, technical terms are included in the text but are "translated" immediately following their use. Of course, all of the geology-related terms used are also defined in the glossary found in the back of this book.

Rocks vs. Minerals

Many people go hunting for rocks and minerals without know-ing the difference between the two. The difference is simple: a **mineral** is formed from the crystallization (solidification) of a chemical compound, or a combination of elements. For example, silicon dioxide, a chemical compound consisting of the elements silicon and oxygen, crystallizes to form quartz, the most abundant mineral on earth. In contrast, a **rock** is a mass of solid material containing a mixture of many different minerals. While pure minerals exhibit very definite and testable charac-teristics, rocks do not and vary greatly because of the various minerals contained within them. This can make identification of rocks far more difficult than minerals.

Important Terms

For those who are entirely new to rock and mineral collecting, there are a few very important terms you should understand not only before you begin collecting, but even before you read this book.

A **crystal** is a solid object with a distinct shape and repeating atomic structure created by the solidification of a chemical compound. In other words, when different elements come together, they form a chemical compound which will take on a very particular shape when it hardens. For example, the mineral galena is lead sulfide, a chemical compound consisting of lead and sulfur, which crystallizes, or solidifies, into the shape of a cube. A "repeating atomic structure" means that when a crystal grows, it builds upon itself. If you have two crystals of galena, one an inch long while the other is a foot long, they would have the same identical cubic shape.

If a mineral is not found in a well-crystallized form but rather as a solid, rough chunk, it is said to be **massive**. If a mineral forms **massively**, it can frequently be found as irregular pieces or masses, rather than as crystals.

Cleavage is the property of a mineral to break in a particular way when carefully struck. As solid as minerals may seem, many have planes of weakness within them derived from the way in which they crystallized. These points of weakness are called **cleavage planes** and it is along these planes that a mineral will separate when struck. For example, galena has cubic cleavage, and even the most irregular, rough piece of galena will break into perfect cubes if very carefully broken.

Finally, **luster** is the intensity with which a mineral reflects light. The luster of a mineral is described by comparing its reflectivity to that of a known material. A mineral with glassy luster, for example, is similar to the "shininess" of window glass. The distinction of a "dull" luster is reserved for the most poorly reflective minerals, while "adamantine" describes the most brilliant. Determining a mineral's luster is something of a subjective experience, so not all observers will necessarily agree.

An Overview of Colorado's Toxic and Radioactive Minerals

Potentially Hazardous Rocks and Minerals

While the vast majority of Colorado's minerals are safe to handle and collect, a few have their own dangers. Potentially hazardous minerals in this book are identified with the symbol shown here; the hazards associated with them are listed here and are discussed in detail in the "notes" section on the pages listed below. Always take proper precautions when handling such materials; see the pages listed for specific precautions and advice.

- Arsenopyrite (page 49)—contains arsenic

- Coloradoite (page 87)—contains mercury

- Galena (page 123)—contains lead

- Serpentine group (page 209)—some varieties are asbestos; thin, flexible crystals which can cause cancer when inhaled

☢ Radioactive Minerals

Radioactive minerals are also included in this book and are marked with the radioactivity symbol shown here. These minerals are listed below, as well as the page number where each can be found:

- Autunite group (page 53)

- Carnotite group (page 69)

- Coffinite (page 85)

- Monazite-(Ce) (page 165)

- Polycrase-(Y) (page 179)

- Samarskite-(Y) (page 203)

- Thorite (page 225)—strongly radioactive

- Uraninite (page 237)—strongly radioactive

- Uranophane (page 239)—can easily embed in skin

- Zippeite (page 249)

- Zircon (page 251)—some, but not all, are radioactive

Radioactive Minerals and Radiation

Colorado is home to many minerals containing uranium or thorium, which are radioactive elements. Radioactive materials can be very dangerous as they emit ionizing radiation, an invisible energy that emanates in all directions from its source and is able to penetrate solid objects. Radiation affects the cells in plants and animals, causing them to mutate and develop cancer. Luckily, most radioactive minerals emit relatively weak radiation and you would have to be in contact with them for extended amounts of time in order to cause permanent harm. Radioactive minerals are prevalent in the state, and due to their beauty and interest, many are sought after by collectors, though great care must be taken when collecting and storing radioactive minerals. Radioactivity is no laughing matter; great harm can come to those who collect irresponsibly.

Radioactivity is detected by an instrument known as a Geiger counter. A Geiger counter is an instrument that will tell you if a mineral is radioactive and how strongly so. When radioactive energy particles contact the sensitive gas-filled tube within the device, a small electrical charge is produced. This charge causes the Geiger counter to produce an audible tone or visual signal informing its user that radioactivity has been detected. For this reason, a Geiger counter is absolutely critical to anyone wishing to collect or protect themselves against radioactive minerals

There are three primary types of radiation: alpha, beta and gamma. Any radioactive mineral can emit any combination of the three types, and not all specimens of a particular mineral will emit the same combination. For example, one mineral specimen may emit only alpha radiation, while another specimen of the same mineral may emit all three types. Alpha, beta and gamma radiation are all dangerous but specimens can be safely stored with proper care and careful collecting:

- Alpha radiation consists of one of the largest and most harmful radiation energy particles, but because of their large size, alpha rays can be stopped by something as thin as a sheet of paper or piece of plastic. You can determine if your specimen is emitting alpha radiation by placing a piece of paper between it and your Geiger counter. If the counter stops reading radiation with the paper in place, you have an alpha-emitting specimen.

- Beta radiation is a smaller particle and requires metal, such as steel or aluminum, to stop the rays. Storing beta minerals in a metal box should stop the majority of the rays. Always double-check with a Geiger counter.

- Gamma radiation is the smallest particle size and it can penetrate almost everything. Strong gamma rays can even go through thin plates of lead, making it one of the most dangerous types of radiation. Thick lead-lined containers are the only way to safeguard against gamma radiation. Due to their pervasive nature, collection of gamma-emitting minerals is not advised.

All three types of radiation particles will naturally lose their energy into the atmosphere and become harmless after they reach a certain distance from the specimen. Alpha particles die out within a few inches from their source, while gamma particles can extend several feet from a specimen. Of course, a Geiger counter will be your best guide to determining how far to keep your distance.

Strength of Radioactivity and Exposure Limits

No matter what type of radiation your specimens may emit, limiting your exposure to them is always important. However, you will inevitably have to handle or transport your radioactive minerals at some point, so it is imperative that you have a sufficient understanding of how much exposure is too much.

Radiation can be measured in several ways. In the United States, the most common method is by counting rems per hour, which measures radiation similar to how a car's speedometer measures miles per hour. "Rem" is short for "roentgen equivalent man," which is a system of measuring radiation's effect on the tissue of a human. This system is named for Wilhelm Röntgen, the physicist who devised the measurement. Since one rem is quite a large amount of radiation, Geiger counters most often count mrem/hour or mR/hr, both short for "millirems per hour" (one millirem is equal to 1/1,000 of a rem). For example, if your Geiger counter says a specimen is emitting 500 mrem/hour, that means it would take one hour for your body to absorb 500 millirems, or two hours to absorb 1,000 millirems, which is equal to one whole rem. To put these measurements into perspective, here are some guidelines for radiation exposure:

- Few people realize that a certain amount of natural radiation enters our bodies on a daily basis. Cosmic rays and radioactive gasses rising from the ground put about 1 mrem of radiation into our bodies every day, averaging around 360 millirems of natural radiation exposure in a year.

- The maximum level of radioactivity allowed in a public area by the government is 48 mrem a day.

- Mineral specimens with radiation counts between 5 and 100 mrem/hour are within the range of relatively safe specimens to collect and store. Most small radioactive specimens are within this range.

- Mineral specimens with radiation counts higher than 100 mrem/hour should only be collected and stored by advanced rock hounds with proper equipment. Minerals in this book that are labelled "strongly radioactive" will often be in this range, even for small specimens.

- Mineral specimens with radiation counts higher than 500 mrem/hour should be avoided by all rock hounds.

Many minerals listed as "strongly radioactive" can be in this range when the specimen is large enough.

- A medical CT scan of your chest gives your body a one-time dose of 800 millirems.

- The maximum safe yearly dose is 5,000 millirems (5 rems). Higher levels of annual radiation greatly increase your risk of developing cancers.

- Symptoms of radiation sickness begin to set in after 50 rems (50,000 millirems) have been absorbed in a short amount of time.

- Exposure to 400 rems (400,000 millirems) and higher can be lethal.

Your body will slowly reduce its level of radiation as your cells repair themselves. However, if you're a careless rock hound, you can easily expose yourself to more radiation than your body can handle. For more information, refer to the United States Nuclear Regulatory Commission's website (www.nrc.gov) or visit the Environmental Protection Agency's website (www.epa.gov).

Precautions for Radioactive Mineral Collecting

It is not recommended that amateurs collect any radioactive minerals. In fact, nobody should plan on starting a collection of radioactive minerals until you are fully prepared to do so by having sufficient knowledge and a safe place to store the specimens. When you feel you are ready, follow these tips carefully:

- If you plan on collecting radioactive minerals, purchasing a Geiger counter is an essential first step. Though they can cost several hundred dollars, a high-quality Geiger counter will not only aid in identifying minerals, but it will help keep you safe. It is not recommended that you collect radioactive minerals without one.

- Ensure that your Geiger counter is calibrated correctly. If it is not, the device may mislead you into believing your specimen is safer than it actually may be. A brand new Geiger counter should be sufficiently calibrated, but older Geiger counters should be checked by a professional calibration service before you begin using it.

- When collecting radioactive minerals, keep only small specimens. The smaller the piece, the less radioactive material it will contain.

- When interacting with radioactive minerals, always wear gloves. Most radiation can still penetrate gloves, but they will keep the harmful materials out of direct contact and eliminate the risk of radioactive dust or mineral fragments becoming embedded in your skin.

- Wear a respirator or dust mask when collecting because many radioactive minerals can dehydrate, causing them to crumble and create a harmful, airborne dust.

- Consider how you'll be storing the specimens. Do not keep them anywhere food or water is stored or where people and animals live or sleep.

- Never keep large amounts. If you do, ensure that there is proper ventilation to outside air in the storage area in order to prevent the build-up of harmful gasses.

- You may also consider purchasing a "pig," which is a lead-lined container designed specifically for the safe storage of radioactive material. Lead is dense enough that it easily blocks alpha and beta radiation. Gamma radiation, however, can only be sufficiently blocked by very thick-walled pigs.

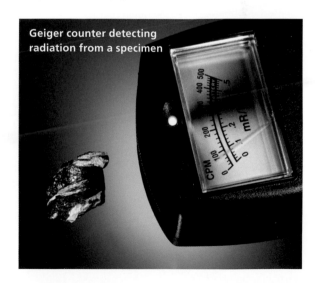

Geiger counter detecting radiation from a specimen

Hardness and Streak

There are two important techniques everyone wishing to identify mineral needs to know: hardness and streak tests. All minerals yield distinct results in both tests, which makes these techniques indispensable to collectors.

The measure of how resistant a mineral is to abrasion is called hardness. The most common hardness scale, called the Mohs hardness scale, ranges from 1 to 10, with 10 being the hardest. An example of a mineral with a hardness of 1 is talc; it is a chalky mineral that you can easily scratch with your fingernail. An example of a mineral with a hardness of 10 is diamond, which is the hardest naturally occurring substance on earth and will scratch every other mineral. Most minerals, including Colorado's, fall somewhere in the range of 2 to 7 on the Mohs hardness scale, so learning how to perform a hardness test (also known as a scratch test) is a key skill to have. Common tools used in a hardness test include your fingernails, a copper coin, a piece of glass, and a steel pocket knife. There are also hardness test kits available that contain a tool of each hardness.

To perform a scratch test, simply scratch a mineral with a tool or another mineral of a known hardness—for example, we know a steel knife has an approximate hardness of 5.5. If the mineral being tested is not scratched, you will then move to a tool of greater hardness until it is scratched. If a tool that is 6.5 in hardness scratches your specimen, but a tool of 5.5 did not, you can conclude that your mineral is about a 6 in hardness. Two tips to consider: as you will be putting a scratch on the specimen, perform the test on the back side of the piece (or, better yet, on a lower-quality specimen of the same mineral), and start with tools softer in hardness and work your way up. On page 18, you'll find a chart that shows which tools will scratch a mineral of a particular hardness.

The second test every amateur geologist and rock collector should know is streak. When a mineral is crushed or powdered, it will have a distinct color—this color is the same as the streak color. When a mineral is rubbed along a streak plate, it will leave behind a powdery stripe of color, called the streak. This is an important test to perform because sometimes the streak color will differ greatly from the color of the mineral itself. Hematite, for example, is a dark, metallic and gray mineral, yet its streak is a rusty red color. Streak plates are sold in some rock and mineral shops, but if you cannot find one, a simple unglazed piece of porcelain from a hardware store will work. There are only two things you need to remember about streak tests: If the mineral is harder than the streak plate, it will not produce a streak and will instead scratch the streak plate itself. Secondly, don't bother testing rocks for streak; they are made up of many different minerals and won't produce a consistent color.

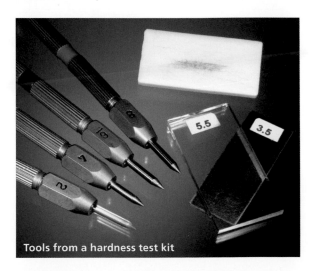

Tools from a hardness test kit

The Mohs Hardness Scale

The Mohs hardness scale is the primary measure of mineral hardness. This scale ranges from 1 to 10, from softest to hardest. Ten minerals commonly associated with the scale are listed here. Some common tools used to determine a mineral's hardness are listed here as well. If a mineral is scratched by a tool, you know it is softer than that tool's hardness.

HARDNESS	EXAMPLE MINERAL	TOOL
1	Talc	
2	Gypsum	
2.5		Fingernail
3	Calcite	
3.5		Copper Coin
4	Fluorite	
5	Apatite	
5.5		Glass, Steel Knife
6	Orthoclase	
6.5		Streak Plate
7	Quartz	
8	Topaz	
9	Corundum	
10	Diamond	

For example, if a mineral is scratched by a copper coin but not your fingernail, you can conclude that its hardness is 3, equal to that of calcite. If a sample is harder than 6.5, or the hardness of a streak plate, it will have no streak and will instead scratch the streak plate itself unless the mineral is weathered or altered by other, softer minerals.

Quick Identification Guide

Use this quick identification guide to help you determine which rock or mineral you may have found. We've listed the primary color groups and some basic characteristics of the rocks and minerals of Colorado, as well as the page number where you can read more about your possible find. The most common traits for each rock or mineral are listed here, but be aware that your specimen can differ greatly.

	If white or colorless and...	**then try...**
	Soft, massive mineral found in chunks rather than crystals	alunite, page 41
	Fairly hard, delicate crystals or masses often found in caves or hot springs	aragonite, page 47
	Flat, bladed glassy crystals that feel very heavy for their size	barite, page 57
	Extremely common soft mineral found as crystal points, masses or veins	calcite, page 67
	Soft, brightly lustrous white crystals or masses found with or on galena	cerussite, page 71
	Very hard botryoidal (grape-like) masses on or in cavities in rock	chalcedony, page 73
	Soft, blocky, square crystals found in sedimentary rocks	dolomite, page 103

WHITE

19

Quick Identification Guide

(continued) **If white or colorless and...**	**then try...**
Common, hard minerals found as blocky crystals embedded in rock	feldspar group, page 109
Very soft, light masses easily scratched by your fingernail	gypsum, page 135
Extremely abundant light-colored rock found in flat, low areas	limestone, page 147
Bright white, soft, coarsely grained rock	marble, page 157
Fairly hard, irregular masses, greatly resembling glass	opal, page 171
Very hard, small and rare crystals with high luster	phenakite, page 175
Very soft masses exhibiting fibrous, fan-shaped structures	pyrophyllite, page 187
Very common, hard mineral forming crystal points, masses or veins	quartz, page 191
Colorless crystals or masses greatly resembling quartz, but much harder	topaz, page 229

WHITE

Quick Identification Guide

(continued) **If white or colorless and...**	**then try...**
Small, delicate crystals formed in cavities in basalt	zeolite group, page 247

If gray and...	**then try...**
Glassy crystals or masses that feel very heavy for their size	barite, page 57
Abundant, dark, fine-grained rock commonly found in western Colorado's mountains	basalt, page 59
Hard, brittle sedimentary rock greatly resembling jasper or quartz	chert, page 77
Delicate, small crystalline "balls" on top of rock	creedite, page 101
Moderately hard hexagonal (six-sided) crystals with flat ends	fluorapatite, page 111
Heavy, dark-colored mineral formed in dense masses or cubic crystals	galena, page 123
Tiny, delicate elongated crystals forming a crust atop rock	hemimorphite, page 139

Quick Identification Guide

(continued) **If gray and...**	**then try...**
Rare, gray-green rock containing large, rounded mineral grains within	kimberlite, page 145
Very abundant, light-colored rock found in flat, low-lying areas	limestone, page 147
Very soft, bluish gray metallic mineral forming coatings on rock	molybdenite, page 163
Rare, metallic gray masses or crude hexagonal (six-sided) crystals	polybasite, page 177
Very hard, grainy rock sharing many of the traits of quartz	quartzite, page 195
Light-colored, fine-grained, and abundant rock found primarily in western Colorado	rhyolite, page 201
Rocks formed in thin layers that can easily be split apart	shale or slate, page 211
Gray metallic masses that bend rather than break	silver, page 215

Quick Identification Guide

	If black and...	then try...
	Dark, elongated crystals embedded in granite	amphibole group, page 45
	Black, radioactive grains contained within sandstone	coffinite, page 85
	Dark, heavy, rectangular crystals embedded in quartz or feldspar	columbite-tantalite, page 89
	Bluish black grains embedded in sandstone	corvusite, page 97
	Small, rectangular, striated (grooved) crystals, often growing with pyrite	enargite, page 105
	Dark, greenish black, coarse-grained rock	gabbro, page 119
	Fibrous, metallic masses or blades with a yellow-brown streak	goethite, page 129
	Shiny, dark, weakly magnetic grains embedded within rock	ilmenite, page 141

BLACK

Quick Identification Guide

continued) If black and...	then try...
Dark, metallic masses that bond strongly with a magnet	magnetite, page 151
Dark, dusty masses or coatings that often leave your hands black	manganese oxides, page 155
Soft masses of thin, layered crystals or grains within rocks	mica group, page 161
Hard, dark masses greatly resembling glass	obsidian, page 167
Dark, heavy, radioactive masses found in pegmatite rock formations	polycrase-(Y), page 179
Glassy, dark grains or masses embedded within dark-colored rocks	pyroxene group, page 189
Hard, glassy, hexagonal (six-sided) crystal points	quartz, page 191
Dense, heavy, shiny mineral found as masses within pegmatite	samarskite-(Y), page 203
Dark, glassy crystals that appear yellow or reddish in bright light and form alongside chalcopyrite	sphalerite, page 219

Quick Identification Guide

continued) **If black and...**   **then try...**

	Small, metallic, triangular crystals	tetrahedrite, page 223
	Long, slender, striated (grooved) crystals embedded in rock	tourmaline, page 231
	Dark, brownish black, dull radioactive masses	uraninite, page 237

If green and...   **then try...**

	Grayish green masses of blocky, elongated crystals	amphibole group, page 45
	Tiny, flat and square radioactive crystals	metatorbernite, page 53
	Deep yellow-green glassy crystals or masses in metamorphic rock	epidote, page 107
	Small, dark, octahedral (eight-faced) crystals embedded in quartz	gahnite, page 121
	Soft, deep green coatings or masses alongside copper minerals	malachite, page 153

BLACK

GREEN

Quick Identification Guide

continued) **If green and...**	**then try...**
Small glassy green grains or crystals embedded in dark rocks	olivine, page 169
Blocky, gray-green masses or grains, especially within dark rocks	pyroxene group, page 189
Very soft, green, greasy-feeling rocks or minerals	serpentine group, page 209
Small "blobs" or coatings of a soft, pale green mineral	smithsonite, page 217
Tiny green grains embedded in sandstone	volborthite, page 243

If blue and...	**then try...**
Thick, blocky, angular crystals in pockets within granite	amazonite, page 43
Soft, vividly colored mineral, often alongside copper or malachite	azurite, page 55
Hard, long, and slender crystals embedded in granite or quartz	beryl, page 61

Quick Identification Guide

BLUE

	continued) **If blue and...**	**then try...**
	Common, soft, brightly colored mineral growing on limonite or copper	chrysocolla, page 81
	Very hard, bluish gray crystals with six sides and steeply pointed ends	corundum, page 95
	Hard, bright masses filling veins and cracks in rock	turquoise, page 234

VICLET OR PINK

	If violet or pink and...	**then try...**
	Soft purple masses filling veins or cavities in rock	fluorite, page 113
	Soft, powdery coatings that rub off on your hands and form on rock surfaces	purpurite, page 181
	Hard, purple, hexagonal (six-sided) crystals tipped with a point	quartz, page 191
	Small, bright pink blocky crystals, often growing with quartz	rhodochrosite, page 197
	Dull, hard pink masses, often with a dusty black coating	rhodonite or pyroxmangite, page 199

Quick Identification Guide

	If red and...	then try...
	Round, hard, glassy crystals or masses, often embedded within rock	garnet, page 125
	Abundant black metallic mineral that turns dusty red on exposed surfaces	hematite, page 137
	Very hard masses that feel smooth and waxy to the touch	jasper, page 143
	Shiny, gray-red crystals or coatings found in silver mine dumps	pyrargyrite, page 183
	Light, grainy rock appearing as if made up of sand	sandstone, page 205
	Brick-red, waxy masses in granite that are very radioactive	thorite, page 225
	Thin, elongated, blade-like crystals	wolframite series, page 245
	Hard octahedral (eight-faced) crystals embedded in granite	zircon, page 251

RED

Quick Identification Guide

ORANGE

If orange and....	then try...
Soft, massive mineral found as rough, irregular masses	alunite, page 41
Abundant, hard grains or masses, especially within granite	feldspar, page 109

YELLOW

If yellow and...	then try...
Fairly hard, glassy crystals or masses, especially in caves or hot springs	aragonite, page 47
Bright yellow crusts of radioactive material	autunite, page 53
Tiny yellow grains of radioactive material embedded within sandstone	carnotite, page 69
Thin, waxy coatings of soft, grayish yellow material on rocks	chlorargyrite, page 79
Moderately hard, glassy six-sided crystals formed in pegmatites	fluorapatite, page 111
Very hard masses that feel smooth and waxy to the touch	jasper, page 143

Quick Identification Guide

continued) **If yellow and...**	**then try...**
Tiny fan-shaped aggregates of needle-like crystals	uranophane, page 239
Radioactive "blobs" or coatings occurring alongside coffinite	zippeite, page 249

If brown and...	**then try...**
Thin, flaky, elongated crystals embedded within quartz or feldspar	astrophyllite, page 51
Soft, abundant crystals filling veins and pockets within sedimentary rocks	calcite, page 67
Hard, dense rocks greatly resembling jasper	chert, page 77
Very fine-grained masses of extremely soft, crumbly material	clay, page 83
Soft, blocky, square crystals found in sedimentary rocks	dolomite, page 103
Sedimentary rocks containing the appearance of plants or animals	fossils, pages 115, 117

Quick Identification Guide

If brown and...   **then try...**

	Rusty brown, dusty, irregular masses that can leave your fingers brown	limonite, page 149
	Groupings of thin, shiny crystals, often arranged into stacks	mica group, page 161
	Reddish brown, crudely shaped radioactive crystals in pegmatite	monazite-(Ce), page 165
	Light brown, dull, crude masses or crystals with black, glassy interiors	polycrase-(Y), page 179
	Light, grainy rock appearing as if made up of sand	sandstone, page 205
	Dark brown, pointed crystals that glow blue under ultraviolet light	scheelite, page 207
	Soft rock that easily splits into thin sheets	shale, page 211
	Soft groupings of thin, rounded, blade-like crystals	siderite, page 213
	Dark brown, glassy wedge-shaped crystal points	titanite, page 227

BROWN

Quick Identification Guide

continued) **If brown and...**   **then try...**

Rare, glassy crystals with square-cross section formed in metamorphic rocks — vesuvianite, page 241

Hard pyramid-shaped crystals embedded in granite or pegmatite rock — zircon, page 251

Tiny triangular crystals embedded in soft white minerals — zunyite, page 253

BROWN

If metallic and...   **then try...**

Silvery crystals with diamond-shaped cross-section — arsenopyrite, page 49

Soft, bronze-brown mineral that develops a blue multicolored tarnish — bornite, page 63

Soft, brittle, and very rare brassy yellow mineral embedded in quartz — calaverite, page 65

Common, soft brassy yellow mineral, often formed with sphalerite — chalcopyrite, page 75

Very rare, soft, silvery mineral, often with multicolored surface tarnish — coloradoite, page 87

METALLIC

Quick Identification Guide

	continued) **If metallic and...**	**then try...**
	Soft, bendable reddish metal	copper, page 93
	Thin, flaky, blade-like crystals, often brassy yellow or metallic blue	covellite, page 99
	Common, heavy, metallic gray mineral forming cubic crystals	galena, page 123
	Metallic black or brown fibrous masses with a yellow-brown streak	goethite, page 129
	Soft, bendable yellow metal	gold, page 131
	Abundant, black metallic mineral with reddish brown streak	hematite, page 137
	Black, metallic grains or crystals that bond strongly with a magnet	magnetite, page 151
	Brassy brown, brittle masses that exhibit deep striations (grooves)	marcasite, page 159
	Very soft, dark bluish gray coatings on rock	molybdenite, page 163

METALLIC

Quick Identification Guide

METALLIC

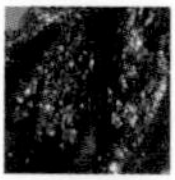	Gray, crudely formed flat hexagonal (six-sided) crystals, often arranged into layered groupings	polybasite, page 177
	Very abundant, brassy yellow cubic crystals with grooved faces	pyrite, page 185
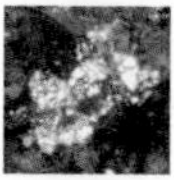	Metallic white, soft, bendable masses, wires, or flakes embedded in rock	silver, page 215
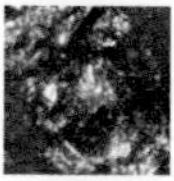	Very rare, brittle, silver colored mineral embedded in rock or quartz	sylvanite, page 221
	Small, black, triangle-shaped crystals	tetrahedrite, page 223

MULTICOLORED

If multicolored or banded and...		**then try...**
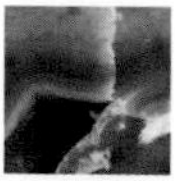	Hard masses of material exhibiting alternating bands of color	agate, page 39
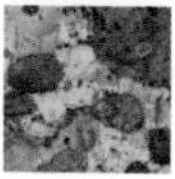	Rock appearing to consist of many smaller pebbles	conglomerate, page 91

Quick Identification Guide

	continued) **If multicolored or banded and...**	**then try...**
	Hard, loosely layered rock, often appearing similar to granite	gneiss, page 127
	Very abundant, hard rock with many large visible grains of different minerals	granite, page 133
	Extremely coarse grained rock, often exhibiting many very large crystals of different minerals	pegmatite, page 173
	Hard, finely layered rock, often consisting of mostly one mineral, such as mica	schist, page 127
	Soft, compact rock containing many small grains of black, glassy material	tuff, page 233

Goethite crystal aggregate

Fibrous, radiating goethite crystals

Massive goethite fragments

Sample Page

HARDNESS: 7 **STREAK:** White

Occurrence

ENVIRONMENT: The type of place where this rock or mineral can be found. For the purposes of this book, the primary environments listed include riverbeds, mountains (primarily in central Colorado), fields (primarily in eastern Colorado), plateaus (flat, high-elevation areas; primarily in western Colorado), quarries (which can include gravel pits), road cuts (where roads have been cut through hills and mountains), and mine dumps (the large piles of waste rock left over at the sites of many old mines).

WHAT TO LOOK FOR: Common or characteristic traits of the rock or mineral

SIZE: The general size range of a rock or mineral

COLOR: The general range of colors a rock or mineral can exhibit

OCCURRENCE: How easy or difficult the rock or mineral is to find. "Very common" means the material takes virtually no effort to find. "Common" means the material can be found with little effort. "Uncommon" means the material may take a good deal of hunting to find. "Rare" means the material will take great lengths of time and energy to find. "Very rare" means the material is so uncommon that you will be lucky to even find a trace of it.

NOTES: This is additional information about the rock or mineral, including more extensive advice about what to look for, specific methods to differentiate it from other rocks or minerals, and interesting facts or unique characteristics

WHERE TO LOOK: Specific geographical areas, ranges, or locations within the state that are good places to start looking for the rock or mineral. Examples are various counties, mining districts, or regions of the state.

Polished agates from near Cañon City

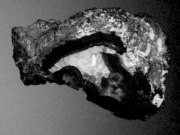

Small agate fragment

Sowbelly agates from near Creede

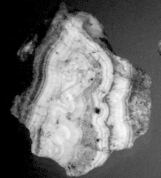

Rough

Cut and polished

Agate

Occurrence

HARDNESS: 7 **STREAK:** White

ENVIRONMENT: Quarries, road cuts, riverbeds

WHAT TO LOOK FOR: Hard, rounded masses of translucent material with a banded interior

SIZE: Agates are rarely larger than an adult's fist

COLOR: White to gray, red, reddish brown, brown to yellow, blue to purple; multicolored banding

OCCURRENCE: Rare

NOTES: Agates are found all over the country and each state produces agates as unique and diverse as the state itself. Agates are a banded variety of chalcedony, which is a form of microcrystalline quartz (quartz crystals too small to see) that often forms within vesicles (gas bubbles) or cavities in rock. The bands of an agate are concentric, which means that they formed one at a time, starting from the outside and growing inward. Therefore, you cannot see the bands when an agate is fully intact; you can only see an agate's beautiful banding if it is broken or cut open, much like an onion's rings. Each band is often colored by various mineral impurities that stain the agate as it is forming. Iron-bearing minerals, for example, frequently turn agates shades of red, orange, brown and yellow, while pure, unstained chalcedony is colorless or white. Because they are made primarily of quartz, agates exhibit all the traits of quartz-based minerals, such as a waxy surface feel and appearance, great hardness, and conchoidal fracture (when struck, circular cracks appear). Colorado has a few varieties of agate scattered throughout the state, but most are quite rare. "Sowbelly agate" is a popular and odd type that has white chalcedony banded throughout purple amethyst.

WHERE TO LOOK: Try looking along the Arkansas River in southern Colorado, as well as in the San Juan Mountains.

Rough alunite

Crudely formed alunite crystals

Alunite

HARDNESS: 3.5–4 **STREAK:** White

Occurrence

ENVIRONMENT: Quarries, mountains, plateaus

WHAT TO LOOK FOR: Irregular masses of light-colored material, often with streaks or coatings of reddish coloration

SIZE: Alunite occurs massively and can be found in any size, from pebbles to boulders

COLOR: White to gray, orange to brown, often with streaks or patches of red or pink color

OCCURRENCE: Common

NOTES: Alunite is a unique mineral that forms when rocks rich with feldspar minerals are altered by sulfur-rich water. This acidic water causes the various minerals in the rock to decompose and reorganize into alunite. Volcanic rocks, particularly rhyolite, are most often affected by this process and the result can be enormous, mountain-sized formations of alunite. Because of the way it forms, this soft mineral is nearly always found massively as rough pieces. Crystals of alunite can be obtained, but they are quite rare and often very crudely formed. Its color can range from white to gray or brown, and flesh-colored streaks or patches are often present throughout a specimen. Identifying alunite can be particularly difficult because of its massive nature and inconsistent coloration. Because of its earthy appearance, it can often resemble limestone, a sedimentary rock that consists primarily of calcite and is very prominent in most western states. Since limestone is easily dissolved in acids, placing a drop of vinegar on the specimen will cause it to effervesce, or fizz, whereas alunite will not be affected. A rough piece of alunite can also resemble a massive specimen of barite, though barite is softer and heavier.

WHERE TO LOOK: Alunite is common in the San Juan Mountains, near the southwest corner of the state.

High-quality, well-formed amazonite crystal cluster

Small greenish crystal

Crudely formed, poorly colored amazonite clusters

Amazonite

HARDNESS: 6–6.5 **STREAK:** White

Occurrence

ENVIRONMENT: Mountains, mine dumps

WHAT TO LOOK FOR: Bluish green stubby, blocky crystals, often alongside smoky quartz

SIZE: Most amazonite crystals range in size from thumbnail- to fist-sized, but rare specimens can be much larger

COLOR: Blue to pale blue, greenish blue, often with white areas

OCCURRENCE: Very rare

NOTES: Amazonite is one of Colorado's best-known collectibles and rock hounds from far and wide visit the state for a shot at collecting their own amazonite specimen. Amazonite is actually a rare bluish green variety of microcline feldspar, which is normally white, pink or orange in color. The source of amazonite's bluish hues had long been a mystery, but recent studies have revealed that the unusual color is a result of lead impurities. Microcline feldspar, including amazonite, forms as coarse, blocky, angular crystals, often intergrown with each other. Many of Colorado's other feldspar minerals form this way as well, but amazonite's color makes it easy to identify by sight alone, rendering any other identification tests unnecessary. The best amazonite specimens are well formed with crisp, sharp angles and have even, bright blue coloration. Amazonite specimens that occur with smoky quartz (gray quartz) crystals are also sought after. Most recent finds, however, are crudely formed, rough-edged crystals. Fantastic specimens are still found each year and make their way into shops where they sell for very high prices, but virtually all come from privately owned mines.

WHERE TO LOOK: Most amazonite localities, especially those in the Pikes Peak area, are on protected or private areas, making specimens nearly impossible for the amateur to find.

Riebeckite

Pegmatite

Massive actinolite

Hornblende

Uralite (gray-green)

Titanite crystal

Amphibole group

HARDNESS: 5–6 **STREAK:** White

ENVIRONMENT: Quarries, mountains, mine dumps

Occurrence

WHAT TO LOOK FOR: Blocky, greenish gray crystals or fibrous, elongated crystals or grains embedded within rock

SIZE: Amphibole group minerals tend to form as thumbnail-sized crystals or in masses smaller than an adult's fist

COLOR: White to yellow, green to gray-green, brown to black

OCCURRENCE: Common

NOTES: Like the pyroxene group, the amphibole group of minerals is an important family of hard, generally dark-colored rock-building minerals. When a mineral is labeled a "rock-builder," it means that they are most commonly found as one of the ingredients within rocks, and in this case, amphiboles are especially common in granite, rhyolite, and impure marble (metamorphosed limestone). Several amphibole minerals are common in Colorado, including tremolite, actinolite, riebeckite and hornblende but primarily occur as black grains embedded in the rocks mentioned above. Actinolite, a magnesium-rich amphibole, is perhaps Colorado's best-known amphibole because of its most unique and collectible form, called uralite. Uralite forms when blocky crystals of diopside, a mineral of the pyroxene group, are replaced by actinolite at a molecular level. The result is called a pseudo-morph, or a mineral that has the outward appearance of a completely different mineral, despite its current chemical composition. Uralite's grayish green color and stubby, square crystals are normally enough to identify it by sight alone. Riebeckite, a black amphibole, can be found as large, coarse, elongated crystals embedded in pegmatites (very coarse granite formations) and granite.

WHERE TO LOOK: Most uralite specimens are found in Chaffee County, while riebeckite is common in El Paso County.

Aragonite stalactites

Broken stalactite showing banded cross-section

Calcite replacement of aragonite

Aragonite crystal aggregate

Aragonite

HARDNESS: 3.5–4 **STREAK:** White

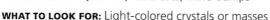

Occurrence

ENVIRONMENT: Quarries, road cuts, mine dumps

WHAT TO LOOK FOR: Light-colored crystals or masses that are harder than calcite and formed within caves or sedimentary rock

SIZE: Aragonite crystals tend to remain smaller than your palm, but massive forms of aragonite can be enormous

COLOR: Colorless, white to gray, yellow to brown

OCCURRENCE: Uncommon

NOTES: Aragonite and calcite have identical chemical composi-tions, yet aragonite is a distinct, separate mineral. This is because of the drastically different crystal structure arago-nite develops as a result of forming in warmer water. This structural difference also makes the two glassy minerals easy to tell apart as it makes aragonite considerably harder, but because aragonite formed under warmer conditions, aragonite is both far less common and stable than calcite. Because of its instability, aragonite actually will, in time, turn into calcite, and this is evident by the hexagonal disc-like crystals found in northern Colorado. These crystals were once aragonite. While aragonite can be found throughout the state in small amounts as veins embedded in rock, the greatest concentrations of aragonite are near hot springs and in limestone caves. However, both environments are dangerous and often protected, so you should not attempt to collect there. To determine if your specimen is aragonite, begin by placing a drop of vinegar on the sample. If it effervesces, or fizzes, you know it is either calcite or aragonite and not a similar mineral. Then, as mentioned above, a hardness test will determine aragonite from calcite.

WHERE TO LOOK: Try Larimer County, in northern Colorado.

Arsenopyrite

Massive arsenopyrite

Diamond-shaped faces

Arsenopyrite crystals

⚠ **Arsenopyrite**

HARDNESS: 5.5–6 **STREAK:** Grayish black

ENVIRONMENT: Mine dumps, mountains

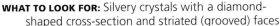

Occurrence

WHAT TO LOOK FOR: Silvery crystals with a diamond-shaped cross-section and striated (grooved) faces

SIZE: Individual arsenopyrite crystals are small—no larger than your thumbnail—but masses or crystal aggregates can be fist-sized or larger

COLOR: Steel-gray to silvery-white

OCCURRENCE: Uncommon

NOTES: Despite their similar names and chemical compositions, arsenopyrite is not actually related to pyrite, primarily because of their different crystal structures. As its name suggests, arsenopyrite contains arsenic, a notoriously toxic element; however, the arsenic molecules are bonded to iron and sulfur molecules, rendering the mineral relatively safe to handle (though you probably will still want to wear gloves). The arsenic does become a danger if the mineral is heated or melted, which will cause the arsenic to be given off as noxious fumes. Arsenopyrite's high arsenic content makes it one of the primary ores of the element, though it has been mined in Colorado for another reason: arsenopyrite often has high amounts of gold and silver impurities. Arsenopyrite is fairly easy to identify due to a few revealing characteristics. One of the biggest clues is its crystal structure, which exhibits a sharp, diamond-shaped cross-section, with striated (grooved) sides. If crystals are not present, a hardness test will differentiate it from pyrite, which is also more yellow in color. Arsenopyrite can match marcasite's hardness, streak, and silvery-gray color, so you'll have to rely on arsenopyrite's most diagnostic trait—when struck, arsenopyrite will release a faint but distinct smell of garlic.

WHERE TO LOOK: Try the mountainous regions in central Colorado.

Astrophyllite crystal aggregate (brownish yellow) in quartz

Astrophyllite (black) in quartz

Large crystal fragment

Astrophyllite

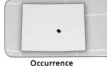
Occurrence

HARDNESS: 3 **STREAK:** Yellowish brown

ENVIRONMENT: Mountains

WHAT TO LOOK FOR: Dark brown, flaky, bladed crystals with high luster embedded in quartz or feldspar

SIZE: Most individual blades of astrophyllite are generally thumbnail-sized

COLOR: Brown to reddish brown, yellow to dark yellow, black

OCCURRENCE: Very rare

NOTES: Astrophyllite got its name from the Greek words for "star" and "leaf," in reference to its thin, flaky crystals that commonly grow in radiating, star-like aggregates. It is a very rare mineral and occurs in very few other places in the world, namely Russia, Greenland and Norway. In Colorado, it can only be found on St. Peters Dome, a mountain southwest of Colorado Springs that contains many pegmatite (very coarse granite) outcroppings. In these rock formations, astrophyllite can be found as crystal aggregates embedded in other minerals. Fine specimens exhibit yellow-brown crystal blades radiating outward from a central point. These dark and reflective layered crystals are easy to spot and identify as they frequently contrast against the light-colored quartz and feldspars they are found within. When crystallized, there is little you could confuse with astrophyllite, though its hardness and flaky crystal habit will help distinguish it. Large, crudely formed masses of astrophyllite can also be recovered, but they lack well-formed crystal shapes and are generally less desirable, despite sometimes being quite large. In pegmatites, it can be found alongside riebeckite, zircon, thorite, quartz and several different feldspars.

WHERE TO LOOK: St. Peters Dome, about 8 miles southwest of Colorado Springs in El Paso County, is the only locality.

Meta-autunite coating

Meta-autunite crystals

Uraninite

Metatorbernite crystals

Metatorbernite crystals (green) on rock

Same specimen as above under short-wave ultraviolet light

Autunite group

HARDNESS: 2–2.5 **STREAK:** Pale yellow

ENVIRONMENT: Plateaus, mountains, mine dumps

Occurrence

WHAT TO LOOK FOR: Yellow, square, flat crystals formed in stacked aggregates

SIZE: Autunite group formations are small and don't normally grow larger than your thumbnail

COLOR: Yellow to greenish yellow, pale to dark green

OCCURRENCE: Uncommon

NOTES: The autunite group is a family of minerals that form as a result of uranium weathering and decomposing, much like the carnotite group. The autunite minerals in general are phosphorus- and uranium-bearing while individual group members contain additional elements like calcium and copper. The two most common autunite group minerals in Colorado's mountainous regions are meta-autunite, which is the partially dehydrated form of autunite, and metatorbernite. Both minerals can still be found in uranium mine dumps as bright yellow or green crusts and coatings atop rock. They can also frequently be found as tiny, square, tabular (flat, plate-like) crystals. In fact, this flat, square crystal structure is the easiest way to distinguish members of the autunite group from other brightly colored radioactive minerals. Meta-autunite is mostly yellow in color and is likely to be found growing on top of black uraninite; the mineral from which autunite is derived. In addition, meta-autunite is fluorescent green under ultraviolet light, a trait which distinguishes it from non-fluorescent metatorbernite. Metatorbernite forms the same square crystals as meta-autunite, but it is green in color. Both minerals have very low hardness, as well as a pearly luster, which is most visible under a bright light.

WHERE TO LOOK: Mine dumps in central and northern Colorado.

Azurite (blue) on rock with malachite (green)

Azurite (blue)

Azurite (blue) intergrown with calcite

Azurite

HARDNESS: 3.5–4 **STREAK:** Light blue

ENVIRONMENT: Mine dumps, mountains, plateaus

Occurrence

WHAT TO LOOK FOR: Richly colored blue crystals or coatings, often alongside green malachite or copper

SIZE: Crystals are rare and smaller than a pea, while masses or coatings can be as large as your palm

COLOR: Dark to light blue

OCCURRENCE: Uncommon

NOTES: Azurite's rich, vivid blue tints have made it one of the world's most sought-after copper-bearing minerals. And while Arizona produces the United States' best specimens, Colorado has yielded many fine examples of the mineral as well. Azurite is very closely related to malachite, another copper mineral that occurs in shades of green, and the two minerals are found intergrown with each other more often than not. This is a result of the fact that the two minerals are chemically identical except for one extra copper molecule in azurite's composition. But that extra copper molecule makes azurite unstable, and over time it will actually lose the excess copper and turn into malachite, changing color in the process. There is little that you could confuse with azurite. Its deep blue color differentiates it from most other common minerals, it nearly always occurs with malachite or copper, and its hardness and streak will always help. Chrysocolla, another soft, bluish copper mineral, is much more common, generally softer than azurite, and is normally a pale blue or bluish green color. It is therefore more likely that you would confuse it with malachite than with azurite. Many Colorado specimens are found in or on calcite, which can also aid in identification.

WHERE TO LOOK: Look in the copper mining districts in the southwest corner of the state, particularly in Ouray County.

Barite crystal aggregates

Clear barite on matrix

Flat crystal grouping

Clusters of crude barite crystals

Barite (Baryte)

Occurrence

ENVIRONMENT: Quarries, road cuts, mine dumps, mountains, fields

WHAT TO LOOK FOR: Light-colored, thin, blade-like crystals that feel very heavy for their size

SIZE: Clusters of barite crystals are generally smaller than a softball, while masses can be very large

COLOR: Colorless to white, gray to grayish blue, brown to reddish

OCCURRENCE: Common

NOTES: Barite, also spelled "baryte," is a common mineral found in many different geological environments. In most places around the country, barite is found as rough, poorly formed veins or masses and uncommonly as fine, collectible crystals. Colorado, however, is particularly privileged because numerous localities throughout the state produce beautiful, well-formed crystals of barite, the likes of which are quite rare in most other states. These crystals range in color from translucent to reddish brown, but few are as attractive or valuable as the pale blue barites found along the Wyoming border. Identifying barite can be difficult, depending on the specimen. If your sample has well-formed crystals, look for the barite to be bladed or tabular (flat, plate-like) in shape, with crystals that are often intergrown nearly parallel to each other, like the pages of an open book. If your specimen is the more common form of barite, which appears as massive veins within rock, rely on other tests. Check its hardness, then feel its "heft." Barite has a high specific gravity, which means that it feels heavy for its size; this is a rare trait among glassy, light-colored minerals.

WHERE TO LOOK: Several areas along the Wyoming border in Weld County have produced the blue barite crystals. Look in the layered sedimentary rock—particularly areas rich with clay.

Basalt

Vesicles (gas bubbles)

Scoria

Basalt with vesicles (gas bubbles)

Basalt

HARDNESS: 5–6 **STREAK:** N/A

ENVIRONMENT: Mountains, road cuts, quarries, mine dumps, riverbeds, plateaus

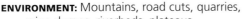

Occurrence

WHAT TO LOOK FOR: Gray to black fine-grained rock, often with many vesicles (gas bubbles)

SIZE: Basalt can be found in any size, from pebbles to boulders

COLOR: Gray to black, reddish brown to brown

OCCURRENCE: Very common

NOTES: Basalt is of the world's most common rocks; it is a dark, very fine-grained rock consisting primarily of the plagioclase feldspars, such as labradorite, and pyroxenes, particularly augite. Other important minerals in basalt include olivine and magnetite, which in high enough concentrations can actually make some specimens of basalt weakly magnetic. It is these dark-colored minerals that give basalt its shades of gray, black and brown. Since basalt is an igneous rock, it is a direct result of volcanic activity and forms when lava, or molten rock, reaches the earth's surface and cools very rapidly. This extremely fast cooling process causes the minerals in the lava to harden before they've grown to a visible size. In contrast, slow-cooling rocks, such as granite, are much more coarse-grained because the molten rock was allowed very long amounts of time to cool, therefore giving its minerals much more time to crystallize. The rapid-cooling nature of basalt also captured gas bubbles that weren't able to escape the lava before it hardened. These rounded cavities are called vesicles and are often the home for future minerals, which form within them. The uppermost portion of a basalt lava flow has many gas bubbles, and when it hardens it makes a lightweight, spongy textured basalt called scoria.

WHERE TO LOOK: Basalt can be found in western Colorado as well as throughout the San Juan Mountains.

Beryl crystals (bluish) in quartz

Crystal cluster

Beryl crystal fragments

Hexagonal crystal cross-section

Aquamarine crystal

Beryl

HARDNESS: 7.5–8 **STREAK:** Colorless

Occurrence

ENVIRONMENT: Mountains, mine dumps

WHAT TO LOOK FOR: Long, slender hexagonal (six-sided), barrel-shaped crystals embedded in rock

SIZE: Beryl crystals are generally an inch or two long and quite thin; however, crystals several feet in size have been found

COLOR: White to gray, pale to dark blue, pale to dark green, pink

OCCURRENCE: Uncommon

NOTES: When most new rock hounds are just getting their start, few are familiar with the mineral beryl—or so they think. Emerald and aquamarine, two famous gemstones that many people know and love, are actually two color variants of beryl that are more common than their sellers would have us believe. In Colorado, both colors can be found, but gorgeous, well-formed aquamarine crystals are so abundant throughout the state that it has been named Colorado's state gemstone. All varieties of beryl form as hard, elongated, hexagonal (six-sided) crystals that are most often found embedded in quartz or pegmatite (very coarse granite formations). These crystals terminate on each end in flattened faces, usually without any kind of distinct termination, or crystal point. This trait distinguishes beryl from other six-sided minerals, such as corundum, quartz or aragonite, though those minerals' hardnesses differ as well. As mentioned above, Colorado beryl is commonly colored shades of blue and green, though other colors like white, gray and pink can be found. Despite its varied coloration, beryl's high hardness, distinct crystal shape, and association with pegmatite formations make it easy to identify.

WHERE TO LOOK: Pegmatite formations in Park, Fremont and Chaffee counties, all in central Colorado, produce beautiful aquamarine crystals embedded in quartz.

Bornite

Multicolored tarnish

Dark blue-gray tarnish

Bornite (brass yellow) in rock

Dark blue-gray tarnish

Bornite

HARDNESS: 3 **STREAK:** Grayish black

Occurrence

ENVIRONMENT: Mine dumps, mountains

WHAT TO LOOK FOR: Bronze metallic mineral, often with a dark blue to purple iridescent surface tarnish

SIZE: Bornite occurs massively and therefore can be found in a wide range of sizes; thumbnail-sized and larger specimens are common

COLOR: Bronze colored, with dark gray to black surface coating exhibiting blue to purple multicolored iridescence

OCCURRENCE: Uncommon

NOTES: Like chalcopyrite, bornite is a combination of copper, iron and sulfur. And while bornite may be less common than chalcopyrite in Colorado, it is by no means rare. It is found in many areas across the state, but particularly in the mountainous central portion of Colorado. It has been used as an ore of copper throughout the United States for decades and has been christened "peacock ore" by miners and rock hounds because of its multicolored iridescent tarnish. This colorful surface coloration develops quickly on a fresh break, turning bornite's usual bronze-brown coloration into flashes of blue and purple. Due to their similar metallic colors, chalcopyrite, pyrite and covellite are minerals you could easily confuse with bornite. Chalcopyrite is slightly harder, and while it can develop a similar tarnish, it is generally found in its original brassy color. Chalcopyrite is also often found crystallized, whereas bornite's cubic crystals are very rare. And it would be hard to confuse bornite with pyrite since pyrite is much harder, doesn't develop a colorful tarnish, is easily found crystallized and is far more common. Bluish bornite can resemble covellite, but covellite is softer.

WHERE TO LOOK: Park County, in the central portion of Colorado, is known not only for bornite, but for very rare bornite crystals.

Calaverite in quartz

Calaverite in quartz

Calaverite crystal
(1mm)

Calaverite

HARDNESS: 2.5–3 **STREAK:** Yellowish gray

Occurrence

ENVIRONMENT: Mountains, mine dumps

WHAT TO LOOK FOR: Small veins or masses of silvery yellow metallic mineral

SIZE: Calaverite occurs as small masses of any size, though specimens larger than your palm are quite rare

COLOR: Silvery metallic white to yellow

OCCURRENCE: Very rare

NOTES: The tellurides are one of the most famous and important mineral groups in Colorado; they are an assortment of minerals that are combinations of tellurium (a rare element used in steel alloys) and rare elements such as gold, silver and mercury. As a gold telluride (a combination of gold and tellurium), calaverite has arguably been the most valuable mineral throughout Colorado's history. Since its discovery, it has been used as an ore of gold and has produced several hundred million dollars' worth of the precious metal. And while it does come from a number of locations in central Colorado, there is no bigger source than the Cripple Creek area in Teller County. Mines in this single locality have produced thousands of pounds of calaverite and are home to the finest specimens in the world. Related telluride minerals, particularly sylvanite, are found there as well. Today, there is considerably less access to calaverite. Privately owned property only complicates the situation for rock hounds—but with persistence and some local help, you should be able to find the metallic yellow mineral embedded in quartz as veins or grains. It can greatly resemble pyrite, a far more common mineral, but a hardness test easily distinguishes the two.

WHERE TO LOOK: Teller County, in central Colorado, and Boulder County, in northern Colorado, have produced the most calaverite.

Steeply pointed crystal

Rhombohedral crystals

Small, ill-formed crystals

Aggregate of crystals

Multi-pointed crystal

Travertine

Aragonite crystals replaced by calcite

Cave calcite

Calcite

HARDNESS: 3 **STREAK:** White

ENVIRONMENT: All environments

Occurrence

WHAT TO LOOK FOR: White crystals, veins or masses that are easily scratched with a copper coin

SIZE: Individual crystals are usually palm-sized or smaller, but masses and veins can be very large

COLOR: Colorless to white, yellow to brown

OCCURRENCE: Very common

NOTES: One of the most common minerals on the planet, calcite is not only easy to find and identify, but it is extremely important for all rock hounds to understand and recognize. Calcite is the primary constituent of limestone, a soft rock that is very abundant in Colorado; calcite is also the primary constituent of travertine, a form of limestone that forms in caves and near hot springs. Calcite often forms beautiful crystals within cavities in limestone and travertine and many other types of rock. These crystals can take dozens of different forms depending on many factors that affect the crystals' formation, such as temperature and pressure. Calcite is commonly found as steep hexagonal (six-sided) crystal points, as rhombohedrons (a shape resembling a leaning cube), or as rounded cave formations. Rarely, calcite is found as rough, flat, hexagonal disks, though these were not originally calcite. They were once aragonite, a mineral that shares calcite's chemical composition but has a different crystal structure. Aragonite's structure is unstable and turns into calcite over time. Calcite can easily be confused with aragonite, or with quartz, another common mineral. Luckily, calcite's hardness is distinctive, a copper coin will scratch it, but your fingernail will not; quartz and aragonite are much harder. Also, a drop of vinegar causes calcite to effervesce.

WHERE TO LOOK: Calcite can be found virtually anywhere.

Carnotite grains (yellow) throughout sandstone

Metatyuyamunite (yellow) on rock

Rock with very rich coating of carnotite

Carnotite group

HARDNESS: 2 **STREAK:** Yellow

ENVIRONMENT: Plateaus, mountains, mine dumps

Occurrence

WHAT TO LOOK FOR: Bright yellow, soft, powdery coatings or stains on or within rock, especially sandstone

SIZE: Individual carnotite crystals are tiny and rare; carnotite coatings or stains can be palm-sized and larger

COLOR: Bright yellow, pale yellow

OCCURRENCE: Uncommon

NOTES: In 1898, Marie Curie, one of the most important figures in the research of radioactivity, discovered the element radium, bringing the world into a new realm of science. A year later, carnotite was first discovered in Colorado and was soon after found to be a source of the elusive element. By 1910, the uranium- and vanadium-bearing mineral deposits of the Uravan Mineral Belt in southwestern Colorado were the world's leading source of radioactive material. Carnotite was the most sought-after mineral in the district and it was mined for decades until larger deposits were discovered in other parts of the world. Today, carnotite group minerals, namely carnotite itself and metatyuyamunite, are found just as they were at the turn of the century. Most commonly, they appear as a yellow stain mixed within sandstone, turning the rock a very bright yellow color. It is also found as thin crusts or coatings atop rock and rarely as tiny crystals. Metatyuyamunite, one of the most common carnotite group minerals in Colorado, is virtually impossible to differentiate from carnotite by the use of simple tests, though the two minerals are so similar in appearance and chemical composition that distinguishing them is often unnecessary.

WHERE TO LOOK: Mine dumps in the Uravan Belt in southwestern Colorado, but be sure to bring your Geiger counter.

Cerussite crystal fragment (½ inch)

Smithsonite (blue-green)

Cerussite (bright white)

Quartz (colorless to gray)

Cerussite

HARDNESS: 3–3.5 **STREAK:** White

ENVIRONMENT: Mountains, mine dumps

Occurrence

WHAT TO LOOK FOR: Small white crystals or dull, heavy masses forming in proximity to galena

SIZE: Cerussite crystals are thumbnail-sized or smaller, while massive cerussite can occur in nearly any size

COLOR: White to gray, yellowish brown, rarely colorless

OCCURRENCE: Uncommon

NOTES: Cerussite is one of the most collectible lead-bearing minerals found in Colorado. Its crystals appear as lustrous, slender, elongated plates that are always small. These heavy, white or gray crystals are often collected for their unique habit of twinned growth. This means that several crystals will often be found intergrown with each other. But while these crystals are the most desirable form of cerussite, most Colorado specimens are very crudely crystallized, if at all. It is much more often found as gray, brown or bright white masses with a dull appearance. Most fragments or masses of cerussite are small and unremarkable and are often completely overlooked by collectors. Cerussite is frequently intergrown with galena, from which it is derived. Galena is the most common and primary lead-based mineral in Colorado. Massive specimens of cerussite can resemble several other minerals, but its high specific gravity (a specimen will feel very heavy for its size) and its association with galena and other lead-based minerals will be a big hint. If you are lucky enough to find cerussite in its crystalline form, there are few minerals with which it can be easily confused. Cerussite crystals are small, mostly opaque, white, very brittle, often brightly lustrous and heavy for their size.

WHERE TO LOOK: Try mine dumps near Leadville in Lake County, as well as dumps in Gunnison and Chaffee counties.

Chalcedony

Chalcedony (white to gray) lining a cavity in rock

Botryoidal (grape-like) chalcedony coating

Chalcedony

HARDNESS: 7 **STREAK:** White

ENVIRONMENT: All environments

Occurrence

WHAT TO LOOK FOR: White to brownish red masses of hard, translucent, waxy-feeling material

SIZE: Chalcedony can occur as nodules (round mineral formations) ranging in size from a pea to a fist or larger; massive specimens can be found any size

COLOR: White to gray, bluish, red to red-brown, yellow-brown

OCCURRENCE: Common

NOTES: Quartz, the most abundant mineral on earth, has many forms, including microcrystalline varieties which develop as hard masses composed of crystals too small to see. Chalcedony is one of the most common of these microcrystalline forms and is found all over the United States. It is frequently found both as rough, irregular masses as well as rounded, smooth botryoidal (grape-like) coatings. And as a quartz mineral, chalcedony exhibits all the traits of quartz, such as high hardness, waxy feel and appearance, and conchoidal fracture (when struck, circular cracks appear). Chalcedony's color varies depending on the impurities found within it. Reds, browns, and yellows are caused by iron, grays and blues are caused by aluminum, and white or colorless chalcedony is pure and unstained. However, nearly all of chalcedony's traits apply to other quartz minerals as well. Jasper, for example, is a quartz mineral that can appear nearly identically to chalcedony except for one trait: jasper is opaque while chalcedony is translucent. Where jasper's quartz microcrystals are compact grains, chalcedony's are tiny parallel fibers which allow light to shine through.

WHERE TO LOOK: Chalcedony can be found virtually anywhere, but specimens are easy to spot in riverbeds and gravel pits.

Chalcopyrite (brass-yellow) on sphalerite (dark gray)

Chalcopyrite crystals (yellow)

Sphalerite (brown)

Tarnished chalcopyrite

Chalcopyrite

Rhodochrosite

Chalcopyrite

Occurrence

HARDNESS: 3.5–4 **STREAK:** Greenish black

ENVIRONMENT: Mountains, plateaus, mine dumps, quarries, road cuts

WHAT TO LOOK FOR: Brittle, golden-yellow crystals, veins or masses that can have a bluish iridescent tarnish

SIZE: Chalcopyrite crystals are thumbnail-sized and smaller while masses or veins can be fist-sized and larger

COLOR: Brassy or golden yellow to brown; sometimes with blue to purple iridescent surface tarnish

OCCURRENCE: Very common

NOTES: Chalcopyrite, like bornite, is a combination of iron, copper and sulfur and has been a major ore of copper for decades. It is present in nearly any mineral deposit in Colorado and is very easy to find and identify. While crystals are the most desirable form, chalcopyrite is generally found massively as brassy yellow veins or grains embedded in rock. When crystals are found, however, they often appear as crude wedge-shaped points. Chalcopyrite's color and luster can be easy to confuse with pyrite, another extremely common sulfur-bearing mineral, though pyrite is harder and a simple hardness test will easily distinguish them. In addition, well-crystallized pyrite is common and is found in cubes— a form which chalcopyrite does not take. Chalcopyrite will occasionally develop a blue or purple iridescent surface tarnish that can greatly resemble bornite's. Telling the two apart can be difficult, but bornite is slightly softer and is much less common. In addition, bornite is almost never found crystallized. Chalcopyrite is also often found with sphalerite, a zinc-bearing mineral that forms in many of the same environments; this is a distinctive association.

WHERE TO LOOK: Literally any mineral deposit throughout the central portion of the state would be a good place to start.

Water-worn chert

Chert fragments

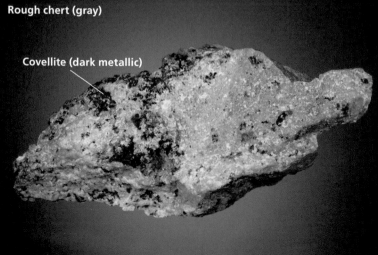

Rough chert (gray)

Covellite (dark metallic)

Chert

HARDNESS: 7 **STREAK:** White

Occurrence

ENVIRONMENT: All environments

WHAT TO LOOK FOR: Very hard, opaque, gray to black or yellow masses

SIZE: Chert occurs massively and can be found in any size

COLOR: White to gray, dark gray to black, yellow to brown

OCCURRENCE: Very common

NOTES: Despite being composed almost entirely of quartz, chert is considered a rock due to the ample mineral impurities and fossil material contained within it. Mineral inclusions are partly responsible for chert's many colors—for example, browns and yellows in chert are caused by iron—but it's the organic fossil matter in chert that gives it its common gray and black coloration. Since most chert formed in bodies of water, microscopic organisms, particularly algae, were fossilized in chert as it formed. And the more fossil material present in chert, the darker the formation becomes. Flint, a black variety of chert famous for its ability to produce a spark when struck, is a particularly fossil-rich variety. Because of this sedimentary process, chert will often have bands of differing color. Since it contains mostly quartz, chert exhibits many traits of the other quartz-based rocks and minerals. For example, chert is very hard and displays a conchoidal fracture pattern (when struck, circular cracks will appear). And, as with most quartz minerals, water-worn specimens of chert will have a waxy look to them, though freshly broken samples will not. Chert is opaque, however, and therefore differs from most quartz-based minerals except jasper. You can differentiate the two by color, since jasper is frequently found in shades of red, green and yellow.

WHERE TO LOOK: Riverbeds are the easiest place to spot specimens.

Chlorargyrite

Chlorargyrite

HARDNESS: 1.5–2.5 **STREAK:** White

Occurrence

ENVIRONMENT: Mountains, mine dumps

WHAT TO LOOK FOR: Translucent greenish yellow coatings on the surface of rocks in silver-bearing regions

SIZE: Chlorargyrite generally forms thin crusts or coatings just millimeters thick, but it can be up to palm-sized in width

COLOR: White to light gray, gray green to yellow; purple-brown when exposed to light

OCCURRENCE: Uncommon

NOTES: Chlorargyrite, a simple mixture of silver and chlorine, is a very soft mineral long used in Colorado as an ore of silver. Despite the fact that it usually occurs in small amounts, some mines, such as those in Custer County, produced so much silver from chlorargyrite that it became the primary ore. Its former name, "cerargyrite," may still be found in older literature, but it refers to the same mineral. Chlorargyrite is white or gray when very freshly formed, but nearly all specimens are found in shades of yellowish gray or greenish gray. If a specimen is exposed to light for extended periods of time, its color will darken to a dull purple-brown color. Perhaps the easiest way to identify chlorargyrite is simply by its appearance. It characteristically appears as dull, soft crusts smeared across the surface of rock. In many ways, it has the look of melted wax, which may be a helpful description for some amateur rock hounds. Chlorargyrite has a particularly low melting point and natural heating in the earth has, in effect, melted most chlorargyrite crystals, giving specimens this "wax on rock" appearance. It can often be found with or on other silver minerals, as well as galena.

WHERE TO LOOK: Mines in Eagle, Lake and Custer counties used chlorargyrite as a silver ore, and specimens can still be found in mine dumps.

Limonite

Chrysocolla

Chrysocolla in limonite

Chrysocolla coating on rock

Chrysocolla

HARDNESS: 2–4 **STREAK:** White to pale blue

ENVIRONMENT: Mountains, plateaus, mine dumps

Occurrence

WHAT TO LOOK FOR: Soft, bright blue-green coatings on copper or masses within rock in copper-rich areas

SIZE: Crusts or masses of chrysocolla generally grow no larger than an adult's fist

COLOR: Pale to dark blue, bluish green, green

OCCURRENCE: Common

NOTES: Chrysocolla is one of the most common and easily identified copper-bearing minerals. Its characteristic bluish green color is indicative of the presence of copper both in the mineral itself as well as the area in which it was found. In fact, for many years chrysocolla served as an indicator of the presence of copper. However, the abundance of chrysocolla does not mean that fine specimens are easy to come by. Crystals of chrysocolla are virtually non-existent and specimens consist of rough, irregular masses embedded within rock or thin, dusty coatings on the surface of material that has been pulled out of copper mines. Pale green coloration, while most common, is far less desirable and valuable than the deep, rich blues found in particularly pure specimens. But with its plentiful nature comes a distinct benefit: as it is a direct result of weathered copper, copper specimens may be nearby, as well as rarer copper-bearing minerals such as malachite or turquoise. However, chryso-colla can be easy to confuse with other massive blue or green copper minerals. Malachite is generally darker green in color, slightly harder, has a fibrous structure, and is less common. Turquoise is much rarer and much harder.

WHERE TO LOOK: Copper is widespread throughout central and western Colorado, particularly Jefferson County, and where you find copper, you'll find chrysocolla.

Montmorillonite

Kaolinite

Bentonite (a variety of montmorillonite)

Clay minerals

HARDNESS: 1–2 **STREAK:** White

ENVIRONMENT: Plains, quarries, plateaus, road cuts

Occurrence

WHAT TO LOOK FOR: Very soft masses of material that easily crumble and have a distinctly chalky or soapy feel

SIZE: The crystals of clay minerals are microscopic; masses of clay can be found in any size

COLOR: White to gray, yellow to brown, greenish

OCCURRENCE: Very common

NOTES: When most people think of clay, they tend to picture wet, sticky mud on a riverbank. However, within that mud are specific minerals in the form of microscopic grains and crystals. These clay minerals fall into two primary groups of clays—the smectite group and the kaolinite group. Both groups are very prevalent throughout the nation, but they are easiest to find as hard masses in the dry western states, including Colorado. Clay minerals can form beds of sedimentary rock hundreds of feet thick, forming hills, plateaus, and rolling plains, so you'll most commonly find them in low-lying flat areas. Montmorillonite, saponite and nontronite are the most common smectite group clays in Colorado, and kaolinite and dickite are the most common kaolinite group clays present. Bentonite is common in Colorado as well and is an impure form of montmorillonite that contains volcanic ash and small amounts of other mixed clay minerals. However, telling any of these clays apart is extremely difficult due to their similar earthy, chalky textures, gray or brown coloration, low hardness and lack of visible crystals. This is true for even advanced rock hounds. Simply being able to identify clay minerals as "clay" can be challenging enough.

WHERE TO LOOK: Clay minerals are especially abundant in the sedimentary-rich plains of eastern Colorado.

Coffinite (black) in sandstone

Layers of higher coffinite concentration

Coffinite (black) with zippeite (yellow)

Uraninite (dull brownish black)

Coffinite (dark shiny black)

 **Coffinite**

HARDNESS: 5–6 **STREAK:** Dark gray to black

ENVIRONMENT: Plateaus, mine dumps

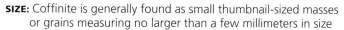

Occurrence

WHAT TO LOOK FOR: Small, black, shiny grains
embedded in sedimentary rock, particularly sandstone

SIZE: Coffinite is generally found as small thumbnail-sized masses
or grains measuring no larger than a few millimeters in size

COLOR: Black, brown

OCCURRENCE: Rare

NOTES: Coffinite and uraninite are the two primary uranium-
bearing minerals from which most other radioactive minerals
in Colorado are derived. Coffinite, the rarer of the two
minerals, is a Colorado type mineral (it was first discovered
in Colorado) and was one of the primary uranium ores in the
uranium- and vanadium-rich Uravan Mineral Belt in southwestern Colorado. This hard, black, radioactive mineral is
rarely found crystallized and instead is collected as massive,
rough pieces. These massive samples are very often
intergrown with uraninite with which it can be easily
confused. Due to their identical hardnesses, you'll have to
rely on other means to differentiate them, such as streak.
Coffinite's streak color will be black whereas uraninite's will
have a brown tint to it. If that doesn't help, uraninite is
generally dull and poorly lustrous, while coffinite is bright,
shiny, and nearly metallic in appearance. But perhaps
coffinite's most common appearance, however, is within
sandstone. In this coarse, grainy rock, coffinite forms as tiny
black masses that fill in the spaces between the sand grains.
This habit will tint the entire rock black, making it difficult to
tell the sand from the coffinite. Only a Geiger counter and a
microscope can help in identifying these specimens.

WHERE TO LOOK: The uranium-rich Uravan Mineral Belt in Mesa,
Montrose and San Miguel counties is the primary locality.

Coloradoite (metallic)

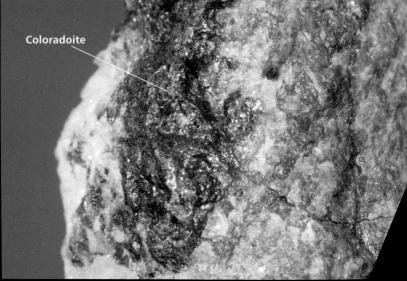

Coloradoite

Coloradoite

Occurrence

HARDNESS: 2.5 **STREAK:** Black

ENVIRONMENT: Mountains, mine dumps

WHAT TO LOOK FOR: Metallic black masses or veins that are brightly reflective and embedded in rock or quartz

SIZE: Coloradoite masses are small, rarely larger than a golf ball

COLOR: Iron-black to metallic gray or silver, sometimes with a gold to blue surface tarnish

OCCURRENCE: Very rare

NOTES: As you may have guessed, coloradoite is named after the state of Colorado, where the first specimen was found. Coloradoite is also a member of Colorado's very famous mineral group known as the tellurides. The tellurides are minerals rich with tellurium, a very rare and valuable element. Colorado has a large amount of these very rare minerals, which are found almost nowhere else in the US. Sylvanite and calaverite are the two tellurides you'll hear the most about because of their gold content and large role in Colorado's mining history; however, despite calaverite and sylvanite's rarity, coloradoite is actually less common. It is only found in a few mines in central Colorado and is extremely difficult to distinguish from the other silver-colored telluride minerals with which it frequently occurs. Coloradoite is slightly softer than calaverite and barely harder than sylvanite, and all of their streak colors differ to a small degree. Luckily, coloradoite often has a gold- or brass-colored surface tarnish with patches of bluish coloration. Coloradoite is a combination of mercury and tellurium, and while it is relatively safe to handle, you'll want to wear gloves and be careful not to inhale any of its dust.

WHERE TO LOOK: Coloradoite was originally discovered in Boulder County, but was later found in very small amounts in Teller and La Plata counties.

Broken rectangular columbite crystal

Embedded columbite crystal

Microcline feldspar

Quartz

Columbite-tantalite fragments

Columbite-Tantalite series

HARDNESS: 6 **STREAK:** Black to dark brown

ENVIRONMENT: Mountains

Occurrence

WHAT TO LOOK FOR: Heavy, black or brown, rectangular-shaped masses in pegmatites (very coarse granite formations)

SIZE: Specimens of columbite-tantalite rarely grow bigger than your thumbnail

COLOR: Black to brownish black

OCCURRENCE: Rare

NOTES: Columbite contains the element niobium (sometimes known as columbium in the US) and tantalite contains the element tantalum. What is significant about these two rare elements, however, is that they are so similar that nothing short of laboratory analysis can tell them apart. This makes columbite and tantalite extremely difficult to differentiate from each other. To make matters more confusing for collectors, each mineral can contain varying amounts of either element. For example, while columbite will, by definition, contain niobium, iron and oxygen, virtually all columbite actually contains a certain percentage of tantalum as well. Minerals that can freely exchange certain elements between each other without changes in their crystal structure are referred to as minerals in series. Therefore, as we cannot always know for sure if a specimen is pure columbite or tantalite, or a mixture of the two, they are simply labelled as columbite-tantalite. In Colorado, they form in pegmatites (very coarse granite formations) as heavy, black rectangular crystals embedded in quartz and feldspars. These crude, dull, stubby crystals or masses can resemble tourmaline, though columbite-tantalite is softer.

WHERE TO LOOK: Try looking in pegmatites in Fremont County.

Conglomerate

Rounded stones

Tuff cement

Conglomerate

Conglomerate

HARDNESS: N/A **STREAK:** N/A

ENVIRONMENT: All environments

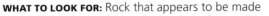
Occurrence

WHAT TO LOOK FOR: Rock that appears to be made
from many smaller rocks that have been cemented together

SIZE: Conglomerate can be found in any size, from pebbles
to boulders

COLOR: Varies greatly

OCCURRENCE: Common

NOTES: Conglomerate is a unique rock formed when small
pebbles or rock fragments are cemented together by
fine-grained sediment. The result is an entirely new variety
of rock, made from the remains of older rocks that have
broken down. Due to its great variability, conglomerate can
be made up of any rock type, depending on where and
when it formed. Conglomerate consists of whole rocks—
particularly rounded pebbles—that have been cemented
together. The cement is often sandstone or tuff, but it can
differ as much as the pebbles contained within it. Conglomerate is most commonly found in sedimentary environments
where the rounded stones accumulated at the bottom of a
lake or river. Finer grains of sediment filled in around the
stones and solidified when pressure forced the material
together. A similar rock type, called breccia, can also be
found throughout Colorado but can be harder to spot
and find. Breccia is formed when rocks are crushed and
broken into jagged fragments before being cemented back
together. The biggest difference between the two rocks is
that the fragments contained within breccia tend to be the
same type of rock, signifying that the original rock was
crushed and cemented all in the same location.

WHERE TO LOOK: Conglomerate is common in sedimentary areas of
both eastern and western Colorado.

Copper

Copper crystals

Copper

HARDNESS: 2.5–3 **STREAK:** Metallic red

ENVIRONMENT: Mountains, plateaus, mine dumps

Occurrence

WHAT TO LOOK FOR: Flexible, reddish metal, often with a black or greenish surface tarnish

SIZE: Copper can occur in a wide range of sizes, from nuggets the size of a pea to sheets several feet across

COLOR: Metallic red or pink to orange, often with a black, green, blue or red surface tarnish

OCCURRENCE: Uncommon

NOTES: Copper, the distinctively reddish metal used for millennia as currency, tools and decoration, is a desirable and sought after collectible for Colorado rock hounds. While not as abundant as in Arizona, copper deposits in Colorado are widespread and were mined for many years. As a native element, copper specimens are found uncombined with any other elements. When copper weathers, however, it merges with other elements and contributes to the formation of many minerals, including malachite, azurite, chrysocolla and turquoise, all of which exhibit characteristic blue and green coloration. These bluish hues are the result when copper weathers and combines with other elements, much the same way that water turns black iron to shades of reddish brown rust. Massive varieties of copper, including nuggets, sheets, and copper-bearing rock, can be found in mine dumps, though after years of collectors sorting through the piles, specimens can be difficult to find. Adding to the difficulty, the soft, malleable (bendable) metal often has a black, gray or greenish surface tarnish, though scratching the specimen should reveal copper's true color below.

WHERE TO LOOK: Copper is widespread throughout central and western Colorado, but large amounts have been found in Jefferson and Montrose counties.

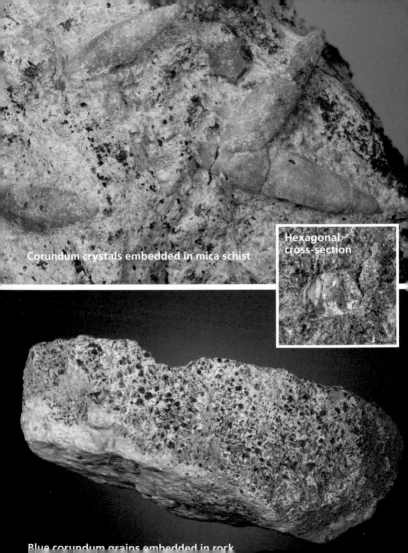

Corundum crystals embedded in mica schist

Hexagonal cross-section

Blue corundum grains embedded in rock

Corundum

HARDNESS: 9 **STREAK:** White

ENVIRONMENT: Mountains, plateaus, mine dumps

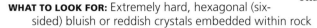

Occurrence

WHAT TO LOOK FOR: Extremely hard, hexagonal (six-sided) bluish or reddish crystals embedded within rock

SIZE: Crystals of corundum tend to remain small, no more than an inch long

COLOR: Colorless, yellow to brown, red, blue to bluish gray

OCCURRENCE: Rare

NOTES: Corundum is a mixture of aluminum and oxygen, and for most collectors, it will be the most easily identified mineral in all of Colorado. That's because only one mineral you'd find in Colorado would be harder—diamond—and they are very rare in Colorado. Indeed, with a hardness of 9, corundum will scratch any other mineral in your collection, including quartz, tourmaline and topaz, all of which are very hard. Due to its great hardness, corundum is widely used as an industrial abrasive; it has also been used in sandpaper, decorative stone and jewelry for centuries. In fact, amateur collectors may know corundum better by its gem names: ruby (for red varieties) and sapphire (for blue, purple and gray varieties). As if its hardness weren't enough, corundum's crystal shape is highly distinctive. It most commonly forms as elongated, slender, hexagonal (six-sided) crystals tipped on both sides with steep points, making the center of the crystal the widest spot. In Colorado, these crystals are frequently found embedded in metamorphic rock, particularly in soft green schists. Rubies have been reported in Colorado, but are rare. Blue and gray sapphires are the primary gem varieties of corundum found in the state.

WHERE TO LOOK: Mountainous Chaffee County produces most of the state's corundum, which is embedded in metamorphic rocks.

Corvusite (black and blue-black) in sandstone

Corvusite (black) on sandstone

Corvusite

HARDNESS: 2.5–3 **STREAK:** Blue-black

ENVIRONMENT: Plateaus, mine dumps

Occurrence

WHAT TO LOOK FOR: Black or bluish black coatings or grains embedded within sandstone in western Colorado

SIZE: Corvusite grains are just one or two millimeters in size, and are found embedded in fragments of sandstone that are generally palm-sized

COLOR: Black, blue-black, brownish black

OCCURRENCE: Rare

NOTES: Corvusite, a vanadium-bearing mineral, is both scarce and difficult to identify. Like coffinite, corvusite is a Colorado type mineral (it was first discovered in Colorado), but that's not the only trait the two minerals share. Both minerals are black, shiny and form primarily as tiny masses in between grains of sandstone. In fact, both minerals can be so prevalent throughout a piece of sandstone that they will often make the entire rock appear black. To make matters more difficult, both minerals form in the same environment—the Uravan Mineral Belt, in southwestern Colorado. But distinguishing corvusite from coffinite can be as simple as testing each with a Geiger counter—coffinite is radioactive while corvusite is not. If you do not have a Geiger counter, look for a bluish tint to the specimen as corvusite will often exhibit a faint, dark blue color. If there are any large grains of the mineral embedded in the sandstone, you can test them for hardness—coffinite is much harder than corvusite. Occasionally, corvusite can be found as a thin coating exhibiting a "slick" or oily appearance on the surface of a piece of sandstone.

WHERE TO LOOK: The Uravan Mineral Belt, in Montrose and San Miguel counties, is the only place to find corvusite.

Clay

Covellite

Chert

Covellite crystals on quartz

Quartz

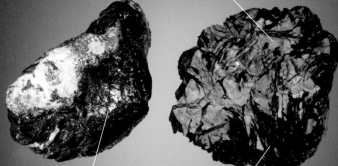

Clay

Covellite (metallic blue)

Bladed covellite crystals (black)

Covellite

HARDNESS: 1.5–2 **STREAK:** Dark gray

Occurrence

ENVIRONMENT: Mountains, mine dumps

WHAT TO LOOK FOR: Thin, metallic crystals often with a brassy or bluish multicolored surface

SIZE: Covellite crystals are rarely larger than your thumbnail while masses can be palm-sized and larger

COLOR: Brassy yellow, blue to purple, rarely red

OCCURRENCE: Rare

NOTES: Covellite, an ore of copper, is easily one of the most attractive metallic minerals you can add to your collection. When covellite's crystals are well developed, they appear as thin, flat, hexagonal (six-sided) disks, though good crystals are very rare. Instead, covellite commonly appears as crude blades or plates embedded in rock, exhibiting shades of brassy yellow, dark blue, or iridescent purple. Many covellite blades form together in layered aggregates which can give specimens a flaky, micaceous look (resembling a mica mineral). In fact, with covellite's extremely low hardness, it could possibly be confused with micas, though this is unlikely as micas tend to be translucent and glassy while covellite is always metallic and opaque. You could, however, confuse covellite with tarnished bornite or chalcopyrite, which can exhibit similar dark blue or purple colorations. A hardness test will tell them apart, as both bornite and chalcopyrite are harder than covellite and neither forms as thin, bladed crystals. As with many other minerals, large specimens of covellite tend to form massively and lack any good crystal structure. Tiny crystals formed within cavities in rocks or other minerals, however, can be nearly perfect.

WHERE TO LOOK: Try mine dumps in the mining districts of the San Juan Mountains in Rio Grande County.

Creedite "balls" (light gray)

Loose creedite crystal aggregate

Creedite coating (light gray) on rock

Creedite

HARDNESS: 4 **STREAK:** White

ENVIRONMENT: Mine dumps

Occurrence

WHAT TO LOOK FOR: Tiny gray-to-purple crystal points formed into ball-like aggregates growing on rock surfaces

SIZE: Individual creedite crystals are just millimeters in size, but crystal aggregates can be thumbnail- to palm-sized

COLOR: Colorless to white or gray, pink to purple

OCCURRENCE: Very rare

NOTES: Like coloradoite, corvusite and carnotite, creedite is a Colorado type mineral, which means that it was first discovered and described in Colorado. It gets its name from the city of Creede in Mineral County, near the location where it was first found. Creedite is a fluorine-bearing mineral that is not only rare in Colorado, but the world over. In fact, Colorado is one of very few places it can be found. Most creedite specimens that make their way into shops come from China and Mexico, while Colorado creedite is virtually unknown to most mineral collectors outside of the state. Individual creedite crystals are very small, elongated prisms that end in a point, but virtually all creedite specimens exhibit hundreds of tiny crystal points arranged into round, ball-like aggregates. These white-to-pink creedite "balls" can be found scattered on the surface of rocks on which they formed. Occasionally, the aggregates grow together and merge to form a thin surface coating on rock. This growth habit is quite unique and it wouldn't be easy to confuse creedite with other minerals. In addition, it is softer than quartz, harder than calcite, and its crystal shape differs from fluorite. Unfortunately for rock hounds, most Colorado creedite comes from deep down in mines, so mine dumps will be your only chance at collecting your own.

WHERE TO LOOK: Try mine dumps in Mineral and Teller counties.

Dolomite crystal aggregates

Dolomite "balls" on quartz

Rhombohedral crystal

Dolomite crystals on chalcopyrite

Dolomite

HARDNESS: 3.5–4 **STREAK:** White

ENVIRONMENT: All environments

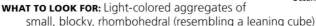
Occurrence

WHAT TO LOOK FOR: Light-colored aggregates of small, blocky, rhombohedral (resembling a leaning cube) crystals, sometimes with slightly curved faces

SIZE: Individual crystals are pea-sized and smaller; crystal aggregates can be palm-sized and larger

COLOR: Colorless, white to gray, pink, yellow to orange, brown

OCCURRENCE: Very common

NOTES: Dolomite is a one of the most abundant minerals in the world, so it's not surprising it's common in Colorado too. Well-crystallized dolomite is common, abundant, and easy to recognize due to its habit of forming rhombohedral crystals (crystals with a shape resembling a leaning cube), often with slightly curved faces. Crystals are often intergrown in large aggregates or crusts, making identification easy. Dolomite forms in a wide range of environments, but particularly within cavities in sedimentary rocks, which produce some of the finest crystal forms. Metamorphic ore-bearing regions, particularly those in central Colorado, produce more complex and interesting specimens, often intergrown with other minerals like tetrahedrite, chalcopyrite and quartz. Despite all its characteristic forms, dolomite is still easily confused with its close mineral cousins calcite and siderite. Both minerals can also form rhombohedrons, as well as be tinted in shades of white or brown. Calcite, however, is softer than dolomite, and siderite is generally crystallized into small, rounded blade or disk-like crystals. If your specimen has no crystals visible, heating a fragment in a flame may help. Siderite will become very weakly magnetic, while dolomite will not.

WHERE TO LOOK: Dolomite is very common throughout the entire state. Try the sedimentary plains of eastern Colorado.

Pyrite (brass yellow)

Enargite crystals

Black enargite veins in rock

Striated crystal (3mm)

Mass of enargite crystals (black) with pyrite (brass yellow)

Enargite

HARDNESS: 3 **STREAK:** Black

ENVIRONMENT: Mountains, mine dumps

Occurrence

WHAT TO LOOK FOR: Metallic black, thick prismatic (elongated in one direction) crystals with grooved faces

SIZE: Enargite crystals are pea-sized or smaller while masses can be palm-sized

COLOR: Steel-gray to grayish black, black

OCCURRENCE: Rare

NOTES: While there once were large deposits of enargite in Colorado, most have long been mined for their copper content. Today, specimens can still be found, but they will take much more work to find. Most enargite is found as black, sooty, indistinct veins embedded in rock. Crystals can be found, however, and are always small, elongated or tabular (flat, plate-like) and have striated (grooved) sides. Despite being a copper mineral, you'll most often find it intergrown with quartz and brassy yellow, iron-bearing pyrite. This relationship can make enargite easy to identify. Poorly formed specimens that do not exhibit the characteristic crystal shape or surface striations can resemble black sphalerite. Enargite is opaque and metallic, whereas even dark specimens of sphalerite are translucent and glassy. In addition, sphalerite is harder.

The mountains of central Colorado, as well as the San Juan Mountains of southwestern Colorado, have produced large amounts of this mineral. Today, some specimens can still be found in mine dumps and in veins running through rock, but it is not as common as it once was.

WHERE TO LOOK: Enargite is widespread throughout central Colorado and significant deposits are in Gilpin, Ouray and Rio Grande counties.

Aggregate of epidote crystals

Very large, well-formed epidote crystal

Epidote crystal

Epidote-rich granite

Allanite-(Ce)

Epidote group

HARDNESS: 6–7 **STREAK:** Colorless to gray

ENVIRONMENT: Mountains, mine dumps

Occurrence

WHAT TO LOOK FOR: Hard green crystals formed in pockets or veins within metamorphic rocks with garnets

SIZE: Most crystals are smaller than a pea, but can rarely grow to thumbnail-sized and larger

COLOR: Light yellow-green to dark green, brown to gray

OCCURRENCE: Epidote is common; allanite-(Ce) is rare

NOTES: Two minerals of the epidote group—epidote and allanite-(Ce)—are found in Colorado. While you're unlikely to confuse the two due to their vastly different geological environments and distinct appearances, both are worthy of mention. Epidote, the mineral from which the group gets its name, is widely collected due to its characteristic color and often well-formed crystals, while allanite-(Ce) is collected due to its rarity and interesting composition. Epidote's trademark yellowish green color is generally enough to identify the mineral. Crystals are common and exhibit two steeply pointed ends with striated (grooved) crystal faces. In addition, epidote is primarily found only in metamorphic rock environments and is abundant in central Colorado's mountains. Pieces of granite can also be colored by epidote, which is caused by chemical alterations changing the rock. Allanite-(Ce) is a very unique kind of mineral known as a rare earth element mineral (REE mineral, for short). REE minerals contain variable amounts of some of the earth's rarest elements. Allanite-(Ce) is rich in the rare element cerium and is only found in pegmatite formations (very coarse granite) as poorly formed, dark gray, elongated crystals that are often embedded within feldspar.

WHERE TO LOOK: Epidote is found in Chaffee and Gunnison counties, while allanite-(Ce) comes from Clear Creek County.

Smoky quartz

Orthoclase feldspar (tan) crystal aggregate

Microcline fragment

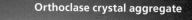

Orthoclase crystal aggregate

Amazonite (blue microcline)

Intergrown orthoclase crystals

Orthoclase crystal

Feldspar group

HARDNESS: 6–6.5 **STREAK:** White

Occurrence

ENVIRONMENT: Mountains, plateaus, mine dumps, quarries, road cuts

WHAT TO LOOK FOR: Very common light-colored blocky crystals or grains within rock, especially granite

SIZE: Feldspar minerals vary greatly in size; crystals range from pea-sized to basketball-sized; massive feldspar specimens often form pea- to coin-sized grains embedded in rock

COLOR: White to gray, cream-colored, flesh-colored, orange, yellow to brown, less commonly blue to green

OCCURRENCE: Very common

NOTES: The feldspars are the most abundant group of minerals on the planet and make up over half of the earth's crust. This extremely important mineral group is one of the primary building blocks of most igneous rocks, especially rhyolite, granite and pegmatite (very coarse granite) formations. The term "feldspar" actually refers to over a dozen distinct minerals, all containing aluminum and silica (quartz material). The feldspars are divided into two subgroups, the potassium feldspars and the plagioclase group. The potassium feldspars include orthoclase and microcline and are common in Colorado. The plagioclase group includes labradorite and anorthite, which are rich in calcium and sodium and make up many dark-colored rocks. Most feldspars are found as grains or masses embedded in rock, particularly granite. When well developed, feldspars tend to form blocky, angular, opaque, light-colored crystals that are often intergrown with each other. Good potassium feldspar specimens are particularly common in Colorado. To identify feldspar, note its abundance and hardness.

WHERE TO LOOK: Feldspars are abundant near Pikes Peak in Teller County and are evident in the orange and pink granite.

109

Dozens of intergrown fluorapatite crystals

Fluorapatite crystal

Tetrahedrite (black)

Gray-blue transparent fluorapatite crystals

Quartz (whit

Fluorapatite

HARDNESS: 5 **STREAK:** White

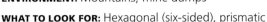
Occurrence

ENVIRONMENT: Mountains, mine dumps

WHAT TO LOOK FOR: Hexagonal (six-sided), prismatic (elongated in one direction) crystals of moderate hardness embedded in rock

SIZE: Colorado's fluorapatite is generally small, with crystals measuring only an inch or two long; massive specimens appear as pea-sized grains embedded in rock

COLOR: Colorless to white or gray, yellow to green, blue, brown

OCCURRENCE: Rare

NOTES: Fluorapatite is the commonest member of the apatite mineral group. It wasn't that long ago, however, that fluorapatite and the rest of the group were simply labeled "apatite." Today, the different varieties of apatite are considered distinct, separate minerals. The apatite group consists of six separate minerals, all of which were named to reflect their composition and their association with the apatite group. For instance, fluorapatite is a fluorine-bearing apatite mineral. As mentioned above, fluorapatite is the most abundant apatite mineral in the world, and in Colorado. Like many minerals, it frequently occurs as embedded grains in other rocks. It often occurs in igneous rocks (such as granite), but it can also be found in sedimentary rocks after it has weathered out of igneous rocks. Fine yellow crystals can rarely be found embedded in Colorado's pegmatite (very coarse granite) formations, while other specimens exhibit intergrown crystals, frequently with quartz. Identifying fluorapatite is fairly easy due to its distinctive hardness—a perfect 5 on the Mohs hardness scale. This means that a steel knife will just barely be hard enough to scratch it.

WHERE TO LOOK: Try pegmatite formations in Eagle County.

Irregular mass of fluorite

Cubic crystal

Massive fragment

Botryoidal (grape-like) fluorite crust

Crude crystal

Tetrahedrite (black)

Fluorite (purple)

Quartz

Fluorite

HARDNESS: 4 **STREAK:** White

ENVIRONMENT: All environments

Occurrence

WHAT TO LOOK FOR: Purple masses or crystals that are harder than calcite, but softer than quartz

SIZE: Fluorite crystals grow no larger than an adult's fist and masses or veins can be nearly any size

COLOR: Colorless to white, purple to blue, yellow to brown, green

OCCURRENCE: Common

NOTES: Fluorite is a common mineral found throughout Colorado in all of the state's different mineral environments. Fluorite crystals can be found in two primary forms in Colorado—glassy cubes or octahedrons (eight-faced crystals). Colorado's fluorite typically exhibits a light to dark purple color, but white, brown and green varieties can be found as well. Massive varieties are often opaque, dark and formed alongside, or intergrown with, other minerals. Veins of the mineral are also common, especially in metamorphic regions, and appear as soft, purple streaks embedded in rock. In the past, some of these veins have been reported to be many feet thick and several miles long. The sedimentary eastern half of Colorado produces fluorite inside pockets within shale and limestone. In the more mountainous regions of western Colorado, fluorite is found as crystals in mine dumps and in pegmatite (very coarse granite) formations. Colorless or white fluorite could be confused with calcite or other similar soft minerals, but identifying fluorite is generally very easy. Its cubic or octahedral crystal habits are highly distinctive, and it is harder than calcite while softer than quartz. And no matter what the specimen's color is, it will always have a white streak.

WHERE TO LOOK: Boulder, Jefferson and Teller counties have produced fine specimens from mining districts.

Fossil insects in
Florissant Shale

Ammonite

Snail shell

Fossils, animals

HARDNESS: N/A **STREAK:** N/A

ENVIRONMENT: Fields, quarries, riverbeds

Occurrence

WHAT TO LOOK FOR: Rocks containing the appearance of animals, shells or bones embedded within

SIZE: Fossils range greatly in size depending on the animal contained within the specimen

COLOR: Varies greatly; fossils take the color of the surrounding rock, primarily shades of gray, yellow and brown

OCCURRENCE: Rare

NOTES: Fossils are formed when the bodies of ancient plants and animals are turned into rock over millions of years. This process begins when an organism dies and becomes covered in sediment, particularly at the bottom of a body of water where it is buried in mud or clay. This heavy silt prevents the creature from being exposed to oxygen and decaying as it normally would on land. With the help of pressure, minerals from the sediment then begin to seep into the once-living cells of the organism and replace its soft tissue with hard rock. As you can imagine, fossils are only found embedded in sedimentary rocks, particularly limestone, sandstone and shale. The Florissant area, west of Colorado Springs, is a particularly well-known fossil site. In Florissant, an ancient eruption of volcanic ash buried a large forested area. Today, the ash has consolidated into shale and between the rock layers you can find hundreds of different insect species. In the limestone beds of eastern Colorado, coral, snail shells, and ammonites (coiled shells of squid-like creatures) can be found. Please note, however, that collecting fossils of vertebrate animals (animals with a backbone), including fish, requires a special permit!

WHERE TO LOOK: While much of the Florissant area is a National Monument, private pay-to-dig quarries are in the area.

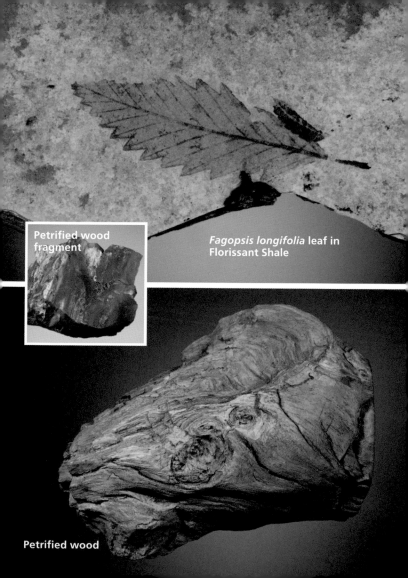

Petrified wood fragment

Fagopsis longifolia leaf in Florissant Shale

Petrified wood

Fossils, plants

HARDNESS: N/A **STREAK:** N/A

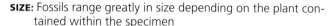
Occurrence

ENVIRONMENT: Fields, quarries, riverbeds

WHAT TO LOOK FOR: Rocks containing the appearance of plants or leaves embedded within

SIZE: Fossils range greatly in size depending on the plant contained within the specimen

COLOR: Varies greatly; fossils take the color of the surrounding rock, primarily shades of gray, yellow and brown

OCCURRENCE: Uncommon

NOTES: While finding a fossilized animal may be a little more exciting, fossilized plants are beautiful and often better-preserved evidence of the ancient past. Plants and animals fossilize in the same way—living organisms are buried in sediment where oxygen cannot reach their tissue, therefore preventing their decay. With pressure, minerals from the sediment begin to replace the once-living cells of the organism and turn it into rock. As a result, fossils are only found in sedimentary rocks like limestone, sandstone and shale, making the sedimentary-rich eastern side of Colorado prime hunting grounds. Petrified wood is one of the more common fossils plants you may find. As its name suggests, it was once wood, and many specimens still exhibit wood grain, growth rings, and knot holes. Many petrified wood specimens are very colorful due to jasper that replaced the tree's cells; others are actually radioactive because of black uraninite, a uranium ore, that made its way into the fossil. In the famous Florissant Shale formation, where volcanic ash buried part of a forest, many different leaves can be found, though the most abundant are from *Fagopsis longifolia*, an ancient species of beech tree.

WHERE TO LOOK: Try the limestone-rich eastern half of Colorado, especially near the Nebraska and Wyoming borders.

Rough gabbro

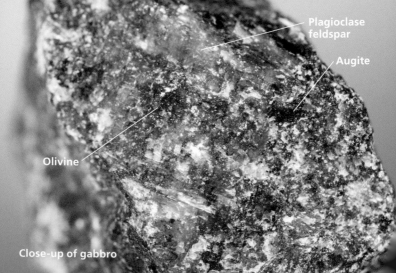

Plagioclase feldspar

Augite

Olivine

Close-up of gabbro

Gabbro

HARDNESS: >5.5 **STREAK:** N/A

ENVIRONMENT: Mountains

Occurrence

WHAT TO LOOK FOR: Dark, coarse-grained rock containing many visible crystals or mineral grains

SIZE: Gabbro is a rock that can form as immense formations, so specimens can be any size

COLOR: Black to gray, greenish, brownish, with mottled appearance

OCCURRENCE: Common

NOTES: The speed at which magma (molten rock) cools determines the size of the mineral grains within it. Grain size is one of the distinguishing features of rocks and signifies where and how the rock formed. For example, basalt's grains are so small that you need a microscope to see them. This is because basalt cooled very rapidly on the surface of the earth. But if the same body of molten rock stays deep within the earth and is allowed a very long time to cool, the individual minerals within the rock have more time to crystallize, thus creating a much larger and more visible grain size. Gabbro actually contains the same minerals as basalt, but it cooled much more slowly. The result is a dark, greenish black rock with coarse, chunky crystal fragments and mineral grains within it. In many specimens, you can easily spot light-colored feldspars, green olivine, and black pyroxenes, like augite. Many of the minerals within gabbro are glassy and reflective, giving the rock a "sparkly" appearance in sunlight. Specimens are not valuable nor very sought after, but they can sometimes contain crudely formed collectible minerals like magnetite and ilmenite.

WHERE TO LOOK: Mountains in north-central Colorado are partially made of gabbro.

Gahnite crystals (green) embedded in quartz

Gahnite crystals (black) on quartz

Gahnite

HARDNESS: 7.5–8 **STREAK:** Gray

Occurrence

ENVIRONMENT: Mountains, mine dumps

WHAT TO LOOK FOR: Hard, dark-colored octahedral (eight-faced) crystals embedded in coarse-grained rock

SIZE: Gahnite crystals are smaller than your thumbnail, but massive and poorly formed specimens can rarely be larger

COLOR: Dark green, dark blue, brown, black

OCCURRENCE: Uncommon

NOTES: Like magnetite, gahnite is a member of the spinel group, a family of minerals known for their sharp octahedral (eight-faced) crystals that resemble two pyramids placed bottom-to-bottom. Gahnite forms primarily within metamorphic rocks like schist and gneiss, but can be found in a few pegmatite (very coarse granite) formations as well. Shades of green are the most desirable colors for gahnite, but black or brown crystals are common. To most collectors, however, color matters little as long as the crystals are well formed with sharp edges and unbroken points. Gahnite crystals are very often embedded in quartz, producing beautifully contrasting specimens. Black gahnite specimens can be very easily confused with magnetite, especially since they can be found in some of the same rock types, but telling them apart is simple. Magnetite is magnetic and is strongly attracted to a magnet whereas gahnite is not. Gahnite is also much harder. Other minerals with similar hardness, like tourmaline and topaz, don't share the same crystal shape. Gahnite does often get overlooked, however, because its crystals are generally very small. Most measure just a few millimeters in width and very rarely get as large as your thumbnail.

WHERE TO LOOK: Fremont County has produced fantastic gahnite specimens both in mine dumps and natural rock outcrops.

Galena (gray) with quartz (white)

Galena crystal fragment

Galena (gray) with rhodochrosite (pink)

Galena (gray) on pyroxmangite (pink)

Galena

HARDNESS: 2.5 **STREAK:** Lead gray

Occurrence

ENVIRONMENT: Mountains, mine dumps, plateaus, quarries

WHAT TO LOOK FOR: Dark-colored, very heavy metallic mineral exhibiting cubic (box-like) structure

SIZE: Galena crystals can range from pea-sized to fist-sized and rarely larger; veins and massive varieties can grow to enormous sizes

COLOR: Dark lead-gray

OCCURRENCE: Common

NOTES: Galena is the primary source of lead, the very soft, extremely dense metallic element known for its toxicity. Because of its rich lead content, galena has a high specific gravity, which means that a specimen will feel very heavy for its size. When well formed, it develops as blocky, cubic crystals, but it most often is found as dark, metallic gray masses or veins embedded in rock. As a simple mixture of lead and sulfur, galena often occurs with other sulfur-bearing minerals, particularly chalcopyrite and pyrite. In Colorado, it is also often found with rhodochrosite and pyroxmangite, two pink manganese minerals, which can be distinctive when one is trying to identify galena in its massive form. With its dark gray color, high density, metallic luster, low hardness, and characteristic associated minerals, galena can be very easy to identify. If you're having trouble, however, there is one more test you can perform. When carefully broken, galena will break into perfect cubes. This is called cleavage, or the action of a mineral breaking along the planes of its molecular crystal structure. This trait will help further distinguish galena from minerals that may share a similar appearance.

WHERE TO LOOK: Try mine dumps in Boulder and Lake counties.

Almandine crystal

Rhyolite

Mass of andradite crystals

Rough almandine crystal

Quartz

Schorl tourmaline

Spessartine garnet mass embedded in pegmatite

Garnet group

HARDNESS: 6.5–7.5 **STREAK:** Colorless

Occurrence

ENVIRONMENT: Mountains, plateaus, mine dumps, quarries, road cuts

WHAT TO LOOK FOR: Very hard, rounded crystals found embedded within rock, especially schist and gneiss

SIZE: Garnets can be pea-sized up to fist-sized, though thumb-nail-sized specimens are most common

COLOR: Deep red to purple, brown, brownish red, greenish, black

OCCURRENCE: Common

NOTES: The garnet group encompasses over a dozen different minerals that all share similar chemical compositions and crystal structures. The majority of garnets, when well developed, form as round, ball-like crystals that exhibit many faces and angles. Many garnets are a product of metamorphism (rocks changed due to heat and pressure); therefore, garnet crystals are often found embedded within metamorphic rocks, particularly schists. Other garnets can be found within gas bubbles in rhyolite or as rough masses within pegmatite (very coarse granite) formations. And after garnets have weathered out of the rock in which they formed, they can be found lying loose in the gravel at the bottoms of rivers. Their hardness, color and crystal shape will help differentiate garnets from other minerals, though distinguishing one type of garnet from another can be very difficult. In Colorado, andradite, almandine, spessartine, grossular and pyrope are the primary garnets collectors can find. They all tend to form in shades of red or brown and are all quite hard, which makes distinguishing them difficult. Doing some extra research will help, but you may have to simply label your specimen as a "garnet."

WHERE TO LOOK: Look in the mountains near Nathrop, in Chaffee County.

Loosely layered granitic gneiss

Granitic gneiss

Garnets in schist

Various schists

Gneiss/Schist

HARDNESS: N/A **STREAK:** N/A

Occurrence

ENVIRONMENT: Mountains, plateaus, mine dumps, road cuts

WHAT TO LOOK FOR: Hard layered or banded rock, often containing pockets of other minerals

SIZE: As metamorphic rocks, gneiss and schist can be found in virtually any size

COLOR: Varies greatly; often multicolored with gray to black, white, green, brown to orange

OCCURRENCE: Very common

NOTES: Metamorphic rocks are those that form as a result of older rocks being changed. Gneiss (pronounced "nice") and schist are two of the primary metamorphic rocks found in Colorado and are hard, layered rocks. These rocks formed when heat and pressure partially melted and compressed pre-existing rocks, arranging the individual minerals contained within them into layers. How tightly spaced these layers are determines the type of rock and how metamorphosed, or changed, the rock has become. Loosely layered, large-grained metamorphic rocks are designated as gneiss, and tightly layered, compact rocks are schist. Both make up a large amount of the Colorado Rocky Mountains and are very abundant. However, the terms "gneiss" and "schist" are simply general labels. To get more specific, we need look to individual specimens. For example, gneisses derived from granite retain some of granite's multicolored and speckled appearance, therefore we call it granitic gneiss. The names of schists are given based on the dominant minerals visible. Mica schist, for example, consists of many layers of shiny mica. Both rocks can contain collectible minerals within, such as garnets, which formed during metamorphosis.

WHERE TO LOOK: Look in the mountains directly west of Denver.

Goethite crystal aggregate

Fibrous, radiating goethite crystals

Massive goethite fragments

Goethite

HARDNESS: 5–5.5 **STREAK:** Yellow-brown

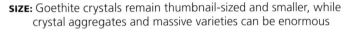

ENVIRONMENT: All environments

WHAT TO LOOK FOR: Metallic brown mineral often found forming as fibrous masses or radiating blades

Occurrence

SIZE: Goethite crystals remain thumbnail-sized and smaller, while crystal aggregates and massive varieties can be enormous

COLOR: Black to brown, yellow to yellowish brown, rusty brown

OCCURRENCE: Very common

NOTES: Like hematite, goethite (pronounced "ger-tite") is an abundant and important ore of iron, long mined all over the world. It shares a number of other traits with hematite as well—both are metallic, have similar hardnesses, form botryoidal (grape-like) crusts, are black to brown in color, and form in the same areas and conditions. However, despite their great similarities, the two iron-bearing minerals can still easily be distinguished from each other. When hematite weathers and oxidizes (combines with oxygen), it turns to shades of reddish brown, but when goethite oxidizes, it turns shades of yellow-brown to orange and rust-colored. These oxidation colors also reflect their different streak colors—goethite streaks yellow-brown, while hematite has a reddish streak. Finally, goethite crystals frequently appear as "fans," or radiating aggregates of needle-like fibers. Goethite is also the primary mineral present in limonite, a mixture of water and various iron oxides. As such, most goethite is found as rusty stains or coatings on the surface of rock or in soil, though collectible varieties are present in Colorado. Fantastic bladed goethite crystals, the likes of which are extremely rare elsewhere in the world, are found in several of Colorado's counties.

WHERE TO LOOK: The Lake George area, in Teller County, has produced some of this country's finest crystallized goethite.

Gold (2mm)

Chalcopyrite

Quartz

Placer gold nugget (5mm)

Calaverite

Gold (yellow)

Quartz

Gold

HARDNESS: 2.5 **STREAK:** Golden yellow

ENVIRONMENT: Mountains, riverbeds, mine dumps

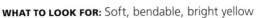
Occurrence

WHAT TO LOOK FOR: Soft, bendable, bright yellow metal embedded in quartz or as grains on river bottoms

SIZE: Gold specimens are normally thumbnail-sized and smaller; gold dust, found in rivers, consists of gold grains that measure just a millimeter or two in size

COLOR: Metallic golden yellow

OCCURRENCE: Very rare

NOTES: Colorado saw its first gold rush in 1858 when thousands of miners flocked to the state for their chance at the legendary metal. Gold is rare, but it is not as impossible to find as most new rock hounds are led to believe—you just need to know how to look for it. Gold primarily forms as veins within rock or quartz where it is both hard to find and difficult to extract, but once erosion and weathering frees the gold from its host material, half of the work is done for you. Freed gold is then found as rounded nuggets at river bottoms. These specimens are called placer (pronounced "plasser") gold and are found this way due to gold's extremely high density, which causes even tiny, pinhead-sized pieces to sink rather than be carried off by the current. Gold's color, low hardness, and density make it extremely easy to identify. Similarly colored minerals, like pyrite, chalcopyrite and calaverite, have more brassy color tones, whereas gold is intensely yellow. All three minerals are also harder than gold (calaverite only slightly so), but with their higher hardness comes brittleness as well. Those minerals will break when struck, but gold will only bend due to its great malleability (bendability).

WHERE TO LOOK: Try mountain streams and rivers in Boulder and Clear Creek counties, but beware of private land.

Granite

Pikes Peak granite

Feldspar (pink, red, orange) masses

Varieties of granite

Diorite

Granite

HARDNESS: N/A **STREAK:** N/A

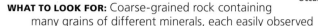

Occurrence

ENVIRONMENT: Mountains, road cuts, riverbeds

WHAT TO LOOK FOR: Coarse-grained rock containing many grains of different minerals, each easily observed

SIZE: Granite is a major rock type found in virtually any size, ranging from pebbles to mountains

COLOR: Varies greatly; multicolored black to gray, white, orange to red or pink, green

OCCURRENCE: Very common

NOTES: A specimen of granite provides the best opportunity to observe the way in which many minerals come together to form a rock. As an igneous rock, it forms from cooling magma, or molten rock. But unlike basalt and rhyolite, two other common igneous rocks, granite cools within the earth, not on the earth's surface. This allows the granite to cool and harden very slowly, which gives the individual minerals contained within the magma enough time to grow to a large, visible size. The result is granite's coarse-grained, mottled appearance where each individual spot of color is a different mineral. The minerals within granite are greatly varied, but potassium feldspars and quartz are the primary components, with plagioclase feldspars, micas, amphiboles and pyroxenes abundant as well. Different proportions of these minerals create different colors of granite. Technically, these different colorations represent distinct, separate rock types, such as diorite, which appears as a dark granite. Though for most collectors, any coarse-grained rock is simply labelled "granite." Colorado is particularly rich with granite, and much of the Rocky Mountains are made up of the coarse rock, especially near Pikes Peak, where it takes on an orange color and can be found along virtually any road.

WHERE TO LOOK: Try looking in the Pikes Peak area on the roadside.

Selenite crystal aggregate

Selenite crystal cluster

Satin spar

Selenite crystal

Alabaster

Satin spar

Gypsum

HARDNESS: 1.5–2 **STREAK:** White

ENVIRONMENT: Fields, plateaus, quarries, road cuts

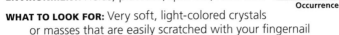

Occurrence

WHAT TO LOOK FOR: Very soft, light-colored crystals or masses that are easily scratched with your fingernail

SIZE: Masses of gypsum can be many feet thick while individual crystals tend to be palm-sized and smaller

COLOR: Colorless to white, gray, yellow to brown, reddish brown

OCCURRENCE: Very common

NOTES: Gypsum is an extremely common and abundant mineral and has been used for decades as the primary ingredient in plaster. This soft, white mineral can be found in nearly all mineral environments because of the different ways it can form. Primarily, it is the result of bodies of water that have dried up and left minerals behind. Therefore, gypsum is very common in the sedimentary regions of Colorado, particularly the low-lying eastern half of the state due to the sea that once covered the area millions of years ago. Another way gypsum forms is from the weathering and decay of sulfur-rich minerals, such as pyrite, so you will also find smaller amounts of gypsum in the metamorphic environments of central Colorado. Normally, gypsum is found as white, chalky masses with no distinct shape, structure or interesting characteristics. This massive variety is sometimes called alabaster. However, two collectible varieties of gypsum exist in Colorado: selenite and satin spar. Selenite is a clear, glassy variety that is often well crystallized. Satin spar consists of many slender parallel fibers, giving it a silky luster. Gypsum can resemble other minerals, but its very low hardness, color, and the fact that it will slowly dissolve in water will help identify it.

WHERE TO LOOK: Gypsum is common everywhere in Colorado; alabaster is found in the east and selenite in the northwest.

Massive hematite with red surface oxidation

Specular hematite

Botryoidal (grape-like) hematite

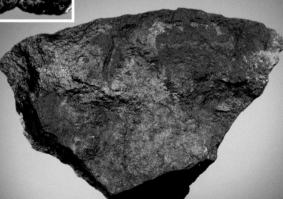

Massive hematite (metallic gray) with red surface oxidation

Hematite

HARDNESS: 5–6 **STREAK:** Reddish brown

ENVIRONMENT: All environments

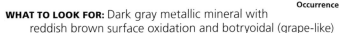
Occurrence

WHAT TO LOOK FOR: Dark gray metallic mineral with reddish brown surface oxidation and botryoidal (grape-like) structure

SIZE: Hematite can occur in very large deposits and some specimens can be boulder-sized; most specimens, however, are palm-sized or smaller

COLOR: White to gray, red, brown, blue to purple

OCCURRENCE: Very common

NOTES: As a simple mixture of iron and oxygen, hematite is the world's most common iron-bearing mineral and has been used as a source of the metal for centuries. In Colorado, it is abundant across the entire state in one form or another. It commonly appears as black, metallic masses or botryoidal (grape-like) coatings on rock, often with a reddish tint, but crystals are extremely rare. In terms of appearance, hematite resembles goethite, a hydrous (water-bearing) iron oxide (combination of iron and oxygen). Despite their similar appearances and hardnesses, they are easy to distinguish. Hematite has a distinctly reddish brown streak color while goethite's streak has a rusty, yellow-brown tint. Hematite may also resemble magnetite or ilmenite, but both of those minerals are magnetic while hematite is not. As mentioned above, there are many ways hematite can form. In addition to its common massive form, a variety called specular hematite, or "specularite," forms in veins, particularly in central Colorado. Specular hematite is formed in higher temperatures and exhibits brightly reflective flakes.

WHERE TO LOOK: Hematite is found literally anywhere. Look for reddish stained rock—hematite is probably nearby.

White hemimorphite in limonite

Massive blue hemimorphite

Aggregate of coarse crystals

Fine hemimorphite crystals

Hemimorphite

HARDNESS: 4.5–5 **STREAK:** Colorless

ENVIRONMENT: Mountains, mine dumps

Occurrence

WHAT TO LOOK FOR: Small, thin and flat glassy gray crystals found in radiating aggregates

SIZE: Individual crystals are generally just millimeters in size and smaller, while crystal aggregates are commonly palm-sized

COLOR: Colorless to white, gray to black, blue to bluish gray

OCCURRENCE: Uncommon

NOTES: Hemimorphite is an important ore of zinc and a popular collectible. It gets its name from the Greek words for "half" and "form," alluding to the fact that hemimorphite's crystals are not symmetrical. Indeed, the small, tabular (flat, plate-like), elongated crystals of hemimorphite are flat on one end and pointed on the other. However, observing this habit is difficult because one of the ends acts as a base from which the crystal grows. Therefore, one of the ends is attached to a matrix, or host rock, which hides the crystal's tip. Hemimorphite frequently forms in radiating aggregates, appearing as "sprays" or "fans" of crystals. Coarse, fibrous masses are also common and consist of many crystals that are tightly grown together. White or gray coloration is common, but rare bright blue varieties exist in Colorado. Well-crystallized hemimorphite is hard to confuse with other minerals, but can resemble certain zeolite minerals. Zeolites, however, only form in cavities within basalt—a rock type hemimorphite rarely, if ever, forms in. Hemimorphite can greatly resemble smithsonite, especially when blue in color, but hemimorphite is slightly harder.

WHERE TO LOOK: Mine dumps near Leadville in Lake County have yielded decent white crystals.

Ilmenite crystal fragment

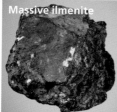

Massive ilmenite

Ilmenite (metallic black) in kimberlite

Ilmenite

HARDNESS: 5–6 **STREAK:** Brownish black

ENVIRONMENT: Mountains, quarries, mine dumps

Occurrence

WHAT TO LOOK FOR: Brittle, black metallic masses
or crystals embedded within rock, often weakly magnetic

SIZE: Embedded grains within rock are smaller than your thumbnail, but massive varieties can be several feet wide in size

COLOR: Metallic black, brownish black

OCCURRENCE: Common

NOTES: Ilmenite, an iron- and titanium-rich mineral, is technically
common, but if you spend some time looking for a nice,
well-formed specimen, you might argue that it is quite rare.
That's because ilmenite is most common as one of the
constituents of many dark-colored rocks, such as basalt and
gabbro, where it only forms as irregular grains. Occasionally,
pegmatite (very coarse granite) formations produce larger
specimens of crudely crystallized ilmenite, but the embedded variety is much more abundant. Of these embedded
ilmenite grains, those found in kimberlite are perhaps the
most attractive. In this occurrence they appear as round,
brightly metallic black grains that contrast against the
greenish rock. If looking through rock doesn't appeal to
you, seek out black sand or gravel, particularly in rivers.
Ilmenite commonly weathers out of rock and accumulates as
pebbles in sedimentary areas. In all occurrences and varieties
of ilmenite, one of its best distinguishing characteristics is its
weak magnetism. This means a magnet will be weakly
attracted to even small, embedded specimens of ilmenite.
Few minerals are magnetic, and the only one you'll confuse
ilmenite with is magnetite. Magnetite is very strongly
magnetic, however, and a magnet will bond firmly with it in
comparison to ilmenite's weak magnetism.

WHERE TO LOOK: Try looking in riverbeds near Estes Park.

Jasper fragments

Multicolored jasper

Jasper

HARDNESS: 7 **STREAK:** White

ENVIRONMENT: All environments

Occurrence

WHAT TO LOOK FOR: Very hard, opaque masses of brown or reddish material, often with a waxy appearance

SIZE: Masses of jasper are generally no larger than an adult's fist, though some specimens can occasionally be much larger

COLOR: Brown, red, yellow to orange, green

OCCURRENCE: Very common

NOTES: Like chalcedony, jasper is a very common form of micro-crystalline quartz (quartz crystals too small to see) that is easily found in any geological environment. It is closely related to chert and therefore forms as opaque, compact masses made up of closely spaced, granular microcrystals. Though chert contains enough impurities to be considered a rock, jasper is pure enough to simply be labeled as a variety of quartz. As such, jasper exhibits all the usual hallmarks of quartz, including a waxy surface feel and luster, great hardness, and conchoidal fracture (when struck, circular cracks appear). Its colors, generally derived from iron impurities, are normally shades of brown, red, yellow and green, which makes it easy to differentiate it from chert's usual white and gray coloration. To distinguish it from chalcedony, another common quartz mineral, look for translucency. Except in very thin specimens, jasper is opaque, and light bounces off its surface rather than passing through it, whereas chalcedony is translucent. This is because chalcedony's microcrystalline structure is fibrous and organized and jasper's is not. Jasper is very easy to find, and road cuts throughout the state often provide easy access to specimens.

WHERE TO LOOK: Jasper is common everywhere, but riverbeds are the easiest place to find beautiful, naturally rounded specimens of hard jasper.

Rough kimberlite

Large pyroxene mass

Finer grained rock

Diamond from kimberlite (3mm)

Rough kimberlite

Kimberlite

HARDNESS: N/A **STREAK:** N/A

ENVIRONMENT: Mountains, plateaus

Occurrence

WHAT TO LOOK FOR: Bluish gray rock containing many large, rounded grains of dark minerals

SIZE: Kimberlite is a variety of rock and therefore specimens can be any size

COLOR: Gray, blue-gray, dark green, black, sometimes yellow

OCCURRENCE: Rare

NOTES: Kimberlite is a rare variety of peridotite, a unique type of rock consisting almost entirely of magnesium- and iron-rich minerals like olivine and pyroxene. Kimberlite is unique because of how it forms. Kimberlite forms deep below the earth's thin crust, in a layer known as the mantle. The mantle is a partially solid body of very hot rock. Pockets of magma (molten rock) form here, and occasionally some of this magma contains large amounts of gasses, such as carbon dioxide and water vapor, which cause the molten rock to boil violently. Eventually, this body of magma will explosively erupt, forcing its way to the earth's surface. The result is a cone-shaped tube of rock, called a kimberlite pipe, that extends from its widest point, at the earth's surface, to its thinnest point, deep in the earth. Kimberlites are scientifically important because they bring material from deep within the earth up to where we can study it, and economically important because kimberlite often contains diamonds. To find kimberlite, look for dark blue or green rocks containing large, rounded grains of dark minerals embedded in a more finely grained mass. Kimberlite easily erodes and alters, forming serpentine minerals in the process. The rock can also have a yellowish tint, caused by decaying iron.

WHERE TO LOOK: Larimer County has a series of kimberlite pipes along the Wyoming border called the State Line Kimberlites.

Travertine

Limestone

Chalk

Dolostone

Limestone

Limestone

HARDNESS: 3–4 **STREAK:** N/A

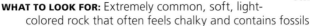

ENVIRONMENT: Fields, plateaus, quarries, road cuts

Occurrence

WHAT TO LOOK FOR: Extremely common, soft, light-colored rock that often feels chalky and contains fossils

SIZE: Limestone formations can form enormous beds and mountains, so specimens can be found in any size

COLOR: White to gray, yellow to brown, red to brown

OCCURRENCE: Very common

NOTES: Limestone is an extremely common sedimentary rock throughout much of the United States, but is particularly in the Great Plains of the central US. And even though Colorado is famous for its Rocky Mountains, nearly half of the state is relatively flat and covered with sedimentary rocks, including limestone. This soft, light-colored rock contains over fifty-percent calcite, along with small amounts of dolomite, clay minerals, and occasionally quartz. When pure, limestone occurs in shades of white or light gray but turns to shades of yellow and brown when iron is present. Its color, low hardness, and abundance make it easy to identify. Because of its high calcite content, it will also fizz in vinegar, which is another distinctive trait. Limestone forms from the remains of aquatic organisms in areas that were once under water. Millions of years ago, a sea connecting the Arctic Ocean with the Gulf of Mexico stretched across North America. When this body of water finally dried up, the flat sea floor hardened and compressed, forming the limestone that underlies much of the Great Plains. Other varieties of limestone include travertine, which forms in caves or hot springs; chalk, a fine-grained white limestone; and dolostone, where dolomite, not calcite, is the primary constituent.

WHERE TO LOOK: The entire eastern half of Colorado is the best place to look.

Massive limonite

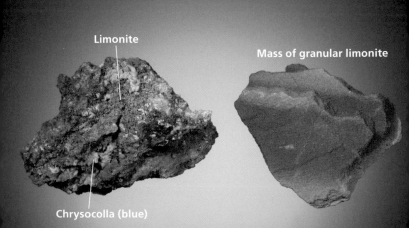

Limonite

Chrysocolla (blue)

Mass of granular limonite

Limonite

HARDNESS: 4–5.5 **STREAK:** Yellowish brown

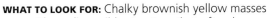

ENVIRONMENT: All environments

Occurrence

WHAT TO LOOK FOR: Chalky brownish yellow masses with no discernible structure that often leave a yellow dust on your hands after handling

SIZE: Limonite occurs massively and can be found in any size

COLOR: Yellow to brown, orange-brown, rust-brown

OCCURRENCE: Very common

NOTES: If you have ever wondered what rust is made of, the answer is limonite. Limonite, however, is not actually a mineral. The name "limonite" is simply a catch-all term that refers to any earthy textured, rust-brown, hydrous (water-bearing) and otherwise unidentified mixtures of iron oxides (combinations of iron and oxygen). Goethite, another iron oxide, nearly fits limonite's description, but it contains less water and has a definite fibrous crystal structure whereas limonite is strictly massive and amorphous (no crystal structure is present at all). Their similarities are not entirely incidental, however; limonite often contains fairly large amounts of goethite in the form of tiny grains. Telling the two apart can be difficult, especially when goethite is massive, but goethite is generally harder than limonite and has the aforementioned fibrous structure. Limonite is also frequently soft and crumbly when weathered and leaves a yellow dust on your hands. It is not particularly collectible and specimens tend to be no more than rough chunks that exhibit the characteristic rusty color. It is also commonly found as patches of reddish brown soil or as brown stains on rock. In fact, limonite is the likely suspect in any rock or mineral that is stained shades of yellow, orange or brown.

WHERE TO LOOK: Limonite can be found virtually anywhere. Mining districts often produce hard, massive pieces of limonite.

Crude magnetite crystals

Calcite

Massive magnetite

Crude loose crystals

Magnetite

HARDNESS: 5.5–6.5 **STREAK:** Black

ENVIRONMENT: Mountains, plateaus, riverbeds, quarries, road cuts, mine dumps

Occurrence

WHAT TO LOOK FOR: Metallic gray mineral that will attract a magnet

SIZE: Magnetite crystals tend to be thumbnail-sized and smaller while massive specimens can be any size

COLOR: Black, iron-gray

OCCURRENCE: Common

NOTES: If you're planning on collecting some samples of magnetite, don't forget a magnet. A magnet will not only help identify magnetite, but it will also aid in finding it. That's because, as its name suggests, magnetite is magnetic and will therefore attract a magnet. Passing a magnet over a specimen of magnetite will cause the two to stick, rendering most other identification tests unnecessary. The only other magnetic mineral easily confused with magnetite is ilmenite, and their similar hardnesses don't help. Ilmenite, however, is only weakly magnetic while magnetite will strongly attract a magnet. Without a magnet, you can easily confuse magnetite with other dark, metallic, iron-bearing minerals like hematite and goethite. Magnetite is most common as embedded grains in dark rocks like basalt or gabbro, but can also be found as crystals within schist or as dark, magnetic grains of sand in rivers. Like gahnite, magnetite is a member of the spinel mineral group. Spinel minerals typically form as sharp octahedral (eight-faced) crystals, a shape resembling two pyramids placed bottom-to-bottom. When these crystals are present and its magnetism has been determined, there is nothing with which you will confuse magnetite.

WHERE TO LOOK: Schists near Estes Park contain crude crystals of magnetite.

151

Malachite (green) with azurite (blue) on calcite (brown)

Malachite coating on rock

Malachite crystal coating

Malachite

HARDNESS: 3.5–4 **STREAK:** Light green

Occurrence

ENVIRONMENT: Mountains, plateaus, mine dumps

WHAT TO LOOK FOR: Richly colored green coatings or fibrous masses alongside copper-based minerals

SIZE: Masses of malachite are found in a wide range of sizes, from pea-sized to basketball-sized

COLOR: Dark to light green

OCCURRENCE: Uncommon

NOTES: While few localities produce malachite as beautiful as Arizona's, Colorado has plenty of the famous copper-based mineral as well. Its deep green coloration makes it both a sought after collectible and easy to identify. It forms as a result of copper weathering and decomposing, so malachite can frequently be found growing on the surface of copper specimens as thin, dusty green coatings. In very copper-rich regions, malachite can be found as thick masses with rounded, botryoidal (grape-like) surfaces and as crusts or coatings on rock surfaces. Crystals are virtually non-existent, but malachite's structure is evident when a massive specimen is broken to reveal its fibrous internal formation. Occasionally, malachite fibers can be seen in undamaged specimens as well. Malachite often forms with azurite, a vividly colored blue copper mineral that is very closely related to malachite. This association can actually help identify both minerals since they occur together so frequently. While telling malachite from azurite may be easy, it can be harder to distinguish from other blue or green copper minerals, particularly chrysocolla. Chrysocolla can form similar dusty green coatings, but is softer, more blue, and generally paler in color than malachite.

WHERE TO LOOK: Try any mine dumps in central and western Colorado.

Manganese coating (black)

Rhodonite (pink)

Dendrites

Psilomelane mass

Wad on pyrite

Manganese oxides

HARDNESS: <6 **STREAK:** Black

ENVIRONMENT: Mountains, plateaus, mine dumps

Occurrence

WHAT TO LOOK FOR: Black, dusty masses or crusts, often formed atop other minerals

SIZE: Manganese oxides can form thin crusts measuring several inches across or dense masses of any size

COLOR: Black, gray, bluish gray, dark brown

OCCURRENCE: Common

NOTES: No group of minerals eludes definitive identification like the manganese oxides. These black minerals range in composition from simple combinations of manganese and oxygen to more complex arrangements of manganese, barium and water. Many manganese oxides form in indistinct masses with no visible crystal structure or identifiable features, and this makes them difficult to identify, as they look alike. This is true for the minerals romanèchite, hollandite and cryptomelane. Determining if a specimen is a manganese oxide is easy, however, as many are black or bluish gray, form powdery coatings or masses atop other minerals, have a sooty black streak, and often leave a black dust on your hands after handling. Instead of assigning a definite label to a specimen, collectors often apply various varietal names to distinguish between unknown manganese minerals. Psilomelane is the name given to hard, dense, metallic masses of manganese minerals that often exhibit a botryoidal (grape-like) surface structure. Thin, dusty coatings or chalky masses are known as wad, and small, branching, tree-like growths on the surface of rocks are called dendrites.

WHERE TO LOOK: Manganese minerals are common in mine dumps as dusty, black soot forming on top of other minerals or rocks.

Yule marble

Yule marble with streaks of silver (black)

Stained marble

Marble

HARDNESS: ~3 **STREAK:** N/A

ENVIRONMENT: Mountains, quarries

Occurrence

WHAT TO LOOK FOR: Soft, white, coarse-grained rock resembling calcite

SIZE: Marble forms massively and can be found in any size, from pebbles to boulders

COLOR: White to gray; black, brown, green or yellow if impure

OCCURRENCE: Uncommon

NOTES: Marble is rather uncommon in Colorado, but one specific variety—Yule marble—has earned itself the title of Colorado's State Stone. Yule marble, named for its discoverer, is a large deposit of nearly pure marble located in the town of Marble (named, of course, for the deposit) in the western-central portion of Colorado. All marble forms when limestone, a soft rock containing over fifty percent calcite, is metamorphosed and changed by heat and pressure. Metamorphosis causes the calcite to recrystallize into a tightly packed mass of tiny crystals, often containing many impurities, such as tiny grains or crystals of dolomite, clay or quartz. Yule marble differs in that it is exceptionally pure, consisting of over ninety-nine percent calcite. The result is a snow-white rock rivaling the quality of the world-famous Carrara marble from Italy. Perhaps the most famous use of Yule marble is in Washington, D.C., where the Lincoln Memorial is built mostly from Colorado's famous rock. While the finest specimens are bright white, there are many samples stained yellow or brown due to iron, green due to serpentine, or black for a number of reasons, including organic material and even silver. If you plan to visit the Marble area, be mindful of your location as the marble is being actively quarried and trespassers are not welcome.

WHERE TO LOOK: The city of Marble is located southwest of Aspen.

Massive marcasite with striated faces

Round marcasite formation

Marcasite crystal in dolomite (5mm tall)

Granular marcasite in quartz (2mm grain size)

Marcasite

HARDNESS: 6–6.5 **STREAK:** Dark gray to black

Occurrence

ENVIRONMENT: Fields, mountains, plateaus, mine dumps, quarries, road cuts

WHAT TO LOOK FOR: Light-colored brassy mineral often formed as tabular (flat, plate-like) crystals or angular masses

SIZE: Individual crystals are generally thumbnail-sized and smaller, but masses of marcasite can be fist-sized and larger

COLOR: Light brass-yellow to brown, silvery white to gray; sometimes with a copper-red tarnish

OCCURRENCE: Common

NOTES: Marcasite is an iron- and sulfur-bearing mineral with the same chemical composition as pyrite, but the conditions in which each form are different, resulting in the two minerals' very different crystal types. Pyrite and its cubic crystals are the most common form taken by this particular chemical composition, so it is much more common than marcasite. Marcasite, however, is still abundant enough for the two minerals to be easily confused. Pyrite is more yellow and brass-like in color, whereas marcasite has a pale silvery brown color. In addition, some specimens of marcasite exhibit a copper-red surface tarnish, making specimens more attractive than usual. Finally, marcasite frequently displays prominent striations (grooves) on its crystals and masses. Pyrite can show these as well, but are less common. In addition, hardness and streak tests help distinguish marcasite from other similar minerals. Unfortunately, marcasite specimens do not survive long in collections. It is an unstable mineral and decomposes easily with exposure to the moisture in the atmosphere, turning to a chalky, gray dust. Specimens may fair better in dry climates.

WHERE TO LOOK: Any ore deposit, especially in central Colorado's mine dumps, is a good place to start.

Biotite mica sheet

Lepidolite mica sheet

Mica crystals in feldspar

Celadonite nodule in basalt

Muscovite mica (light brown) with fuschite (green muscovite)

Mica group

HARDNESS: 2.5–3 **STREAK:** Colorless

ENVIRONMENT: All environments

Occurrence

WHAT TO LOOK FOR: Highly reflective, dark-colored minerals that form in stacks of thin, flexible sheets

SIZE: Most flakes of mica are thumbnail-sized or smaller, but they can grow to palm-sized or larger in coarse granite formations

COLOR: Colorless, brown, gray to black, less commonly purple, pink, green, and bluish green

OCCURRENCE: Very common to uncommon, depending on variety

NOTES: The mica group is a family of minerals containing several dozen members, all with similar crystal structures and compositions. They all tend to form as thin, flexible, sheet-like crystals arranged into layered stacks that resemble the pages of a book. Most are also very lustrous and reflective. Therefore, mica minerals can generally be easily identified as a mica, but individual mica minerals can be very difficult to distinguish from each other. In Colorado, many micas are present throughout the state, but luckily, several of them are easy to identify. Muscovite is one of the most common and it is generally brown to white in color, though a rarer green chromium-bearing variety, called fuschite, is popular with collectors. Biotite is somewhat more rare and is very dark or black in color. Both biotite and muscovite are common as the dark, reflective spots in rocks, primarily granite. Lepidolite is a very collectible (but uncommon) lilac-purple mica, which is colored by lithium. Celadonite is different in that it rarely displays the usual layered crystals of mica. Instead, it forms bluish green coatings and nodules (round mineral clusters) on and in rock. Celadonite is easily identified by visual means alone; its color is very distinctive.

WHERE TO LOOK: Look amid pegmatites and granites in central Colorado.

Metallic molybdenite coating on granite

Molybdenite crystal in quartz

Molybdenite vein in quartz

Molybdenite

HARDNESS: 1–1.5 **STREAK:** Grayish green to black

Occurrence

ENVIRONMENT: Mountains, mine dumps

WHAT TO LOOK FOR: Small, very soft, metallic flakes, crystals or veins embedded within quartz or rock

SIZE: Individual molybdenite crystals are normally no larger than a pea, but masses or aggregates can be up to fist-sized

COLOR: Metallic blue-gray

OCCURRENCE: Common

NOTES: Molybdenite is the primary source of the element molybdenum, which is used in the production of steel alloys and chrome. Colorado was once one of the world's leading producers of the element, especially from mines in Lake County. When well formed, molybdenite develops as small, hexagonal (six-sided) crystals that are very thin and flat. These crystals are often arranged into stacks, creating layered aggregates much like those of mica. However, in Colorado, well-developed crystals are quite rare and molybdenite is primarily found as thin coatings on granite or as veins throughout other rocks, formed within fissures and cracks in the stone. Even without visible crystal structure, however, molybdenite is very easy to identify. It is always dark bluish gray and has a bright, metallic luster. While other minerals may share a similar appearance with molybdenite, few are as soft. In fact, molybdenites extreme lack of hardness is the most important factor to note when identifying it, as it will narrow down the possibilities very quickly. However, if in doubt, there is yet one more characteristic trait of molybdenite: on a streak plate, its streak color is grayish green, but on paper, it is black.

WHERE TO LOOK: Try mine dumps in Lake County, particularly near Leadville where molybdenite was mined for years.

Monazite-(Ce) crystal cluster

Crude crystal points

Monazite fragment

Crude crystal face

 **Monazite-(Ce)**

HARDNESS: 5–5.5 **STREAK:** Nearly white

ENVIRONMENT: Mountains, mine dumps, riverbeds

Occurrence

WHAT TO LOOK FOR: Rough, poorly formed, reddish brown prismatic (elongated in one direction) crystals embedded within pegmatites (very coarse granite formations)

SIZE: Monazite-(Ce) crystals are rarely longer than an inch or two; grains found in sand measure just a few millimeters in size

COLOR: Dark red, reddish brown, brown to yellow, dark orange

OCCURRENCE: Uncommon

NOTES: Monazite-(Ce) is one of the few minerals that contains rare earth elements. Such minerals are called REE minerals, for short. REE minerals contain variable amounts of some of the earth's rarest elements, such as yttrium and cerium. In REE minerals, many rare earth elements may be present, but the element in highest quantity is noted in parentheses after the mineral's name. Therefore, in monazite-(Ce), cerium is the most abundant rare earth element found in this variety of monazite. Monazite-(Ce) is one of the most common REE minerals in the world and has long been mined. While it can be found as tiny grains within igneous and metamorphic rocks, it is most easily found and identified with pegmatite (very coarse granite) formations. In pegmatites, monazite-(Ce) is found as rough, opaque, elongated crystal points in shades of brown, red or yellow. These crystals are often cracked and easily break into fragments if you're not careful. Once it weathers out of the rock in which it grows, monazite-(Ce) makes its way to riverbeds and shores in the form of sand. But no matter how you find it, monazite-(Ce) nearly always has ample amounts of thorium in its composition, making the mineral strongly radioactive.

WHERE TO LOOK: Pegmatites in central Colorado, especially around Pike's Peak, are the best source of crystallized monazite-(Ce).

165

Obsidian fragment

Conchoidal fracture

"Apache tears"

Perlite matrix

Obsidian

HARDNESS: 6–7 **STREAK:** N/A

Occurrence

ENVIRONMENT: Mountains, plateaus

WHAT TO LOOK FOR: Dark, black, translucent rock with a distinctly glass-like appearance

SIZE: Obsidian can occur in any size, but it generally is found fist-sized or smaller

COLOR: Black, dark brown, gray

OCCURRENCE: Uncommon

NOTES: The first time you find a shard of obsidian, you may discard it as simply a piece of dark glass. Actually, it is glass, but it is volcanic glass, created during an eruption of lava (molten rock). Like rhyolite or basalt, it was spewed onto the earth's surface where it was allowed to cool and harden, but unlike other volcanic rocks, obsidian hardened very rapidly when it contacted very cool air or water. This extremely fast cooling allowed none of the minerals within the lava enough time to crystallize and they were instead frozen in place as microscopic grains. The result is a natural glass, generally black or dark brown in color and translucent when in thin pieces. Obsidian contains large amounts of quartz and feldspars, so if it were allowed more time to cool on the earth's surface, it would have formed rhyolite. Obsidian's distinctive dark, glassy appearance is normally enough to identify it. If there is any doubt, there are a few other traits to look for. It is very hard and brittle, much as you would expect glass to be, and has very prominent conchoidal fracture (when struck, circular cracks appear), apparent in nearly all specimens as sharp edges. "Apache tears" is the name collectors apply to round masses of obsidian that form within perlite, which is altered, water-bearing obsidian.

WHERE TO LOOK: Try looking in western Colorado's San Juan Mountains.

Olivine crystals embedded in basalt

Peridot (forsterite) fragment

Gabbro fragment

Olivine (greenish areas)

Olivine group

HARDNESS: 6.5–7 **STREAK:** Colorless

ENVIRONMENT: Mountains, plateaus, mine dumps, quarries, road cuts

Occurrence

WHAT TO LOOK FOR: Hard, green, translucent grains embedded within coarse-grained rock, especially gabbro

SIZE: Most grains or pockets of olivine in rock measure smaller than an inch, though larger, rarer masses have been found

COLOR: Yellowish green, pale to dark green, brown-green, brown

OCCURRENCE: Forsterite is common; fayalite is rare

NOTES: While the olivine group contains several minerals, only two are prominent in Colorado. Forsterite is a magnesium-rich olivine mineral that appears in shades of yellowish green to dark green. Fayalite, an iron-bearing olivine, is found in shades of brownish green to dark brown. Of the two, forsterite is much more common and it is likely the only olivine mineral you will easily find. Both primarily form as hard, glassy grains embedded in dark-colored rocks; they rarely form as well-developed crystals. In fact, the easiest way to find forsterite and other olivines is to closely inspect some samples of gabbro, a type of dark, coarse-grained rock; the translucent, dark green-yellow grains and masses embedded in the rock are olivine. Due to their identical hardnesses, the two olivine minerals can be very difficult to distinguish when their colors are similar. In addition, laboratory testing is usually required to determine the identities of specific olivine minerals, so labelling a mineral as an olivine is as much as most collectors can do. Light green, translucent samples of olivine, particularly forsterite, are known as peridot, which has been used as a gemstone for centuries. Peridot is rare, but can form in basalt or kimberlite.

WHERE TO LOOK: The volcanic rocks of the San Juan Mountains, in southwestern Colorado, can contain peridot.

Common opal (glassy white)

Green "fire" in opal

Calcite crystal points with intergrown hyalite fluorescing green under ultraviolet light (this specimen does not show hyalite in normal light)

Opal

HARDNESS: 5.5–6.5 **STREAK:** White

Occurrence

ENVIRONMENT: Mountains, plateaus, road cuts

WHAT TO LOOK FOR: Smooth masses of light-colored material that greatly resemble colored glass

SIZE: Opal forms massively and can be found in chunks that are generally smaller than an adult's fist

COLOR: Colorless, white to gray, yellow to brown, blue, pink

OCCURRENCE: Common

NOTES: Opal is a popular collectible that comes in several varieties and appearances. However, a specimen of opal can actually be one of several different materials. All opals are glassy masses consisting primarily of water and silica (quartz material), but how the silica molecules are arranged can vary. In cristobalite, one of the most common opal varieties, there is a partial order to the silica molecules. The silica is arranged into microscopic stacks of tiny spheres, while the outward appearance of the opal remains massive and without any particular shape. Cristobalite is typically referred to as "common opal" because of its solid, opaque appearance and white, gray or brownish colors. Another variety of opal, simply called "opal-A," is an opal with less organization to its silica molecules and ample water in its composition. Opal-A is also the most likely variety to exhibit opal's famous "fire," which are areas of brightly iridescent color. Opal's fire is caused by light bouncing between microscopic layers in opal's structure. Finally, hyalite is a form of silica glass, with no internal structure, and it is transparent and often fluorescent. All opals can crack and crumble in collections due to their water content evaporating. Many opal collectors keep their specimens in jars of mineral oil to help retain their water content.

WHERE TO LOOK: Try looking in volcanic rocks in western Colorado.

Pegmatite

Feldspar (tan)

Quartz (white)

Schorl (black)

Muscovite (gray)

Riebeckite crystal

Pegmatite

Quartz (gray)

Schorl

Feldspar (white)

Muscovite

Spessartine (red)

Pegmatite

HARDNESS: N/A **STREAK:** N/A

ENVIRONMENT: Mountains, quarries

Occurrence

WHAT TO LOOK FOR: Extremely coarse-grained rock comprised of large, well-formed crystals of various minerals

SIZE: Pegmatite formations can be huge, so specimens of the rock can be found in virtually any size

COLOR: Varies greatly; multicolored primarily with white to gray, black, pink, yellow to brown, green

OCCURRENCE: Uncommon

NOTES: Pegmatite is a unique variety of rock that forms in the lowest portion of a granite formation, deep within the earth where it is very hot. Hotter temperatures prevent magma (molten rock) from cooling quickly, which allows the minerals contained in the rock more time to crystallize to a visible size. While the slow-cooling granite seated above pegmatite may exhibit coarse, pea-sized mineral grains, pegmatite cools so slowly that entire crystals, often several inches in size, are allowed to fully form. As you might guess, pegmatite contains many of the same minerals as granite, including feldspars, micas and amphiboles, as well as plenty of quartz, all crystallized to a large, collectible size. Pegmatite forms so deep down, however, that other, rarer minerals can be present, including precious metals, like gold, and rare earth element minerals, such as monazite-(Ce) and samarskite-(Y). For the same reason, pegmatites also often contain many radioactive elements. Pegmatite formations are a rock hound's dream—the minerals contained within them are often large, well formed, rare, and fairly easy to separate from the rock. Pegmatite outcroppings are scattered throughout northern and central Colorado, but can be hard to spot when grown-over with plants.

WHERE TO LOOK: Pegmatites are abundant around Pikes Peak.

Crust of colorless phenakite crystals on rock

Phenakite crystals

White phenakite crystal

Phenakite

HARDNESS: 7.5–8   **STREAK:** White

ENVIRONMENT: Mountains

Occurrence

WHAT TO LOOK FOR: Small, very hard, white crystals
that are often very well formed and found in pegmatites
(very coarse granite formations)

SIZE: Phenakite crystals tend to be small, growing no larger than
a golf ball

COLOR: Colorless to white, gray, yellow, pale pink

OCCURRENCE: Rare

NOTES: Phenakite is one of Colorado's rarer minerals, but one
for which the state is well known. It gets its name from
the Greek word meaning "deceiver," due to its similar
appearance to quartz. Phenakite, however, is much harder
than quartz and most other minerals; its considerable
hardness is due to its beryllium content. It forms two distinct
crystal shapes in Colorado, depending on which of the two
major phenakite localities a specimen came from. In the
Crystal Park area, near Colorado Springs in El Paso County,
phenakite is found as rhombohedrons (a shape resembling
a leaning cube) alongside amazonite. In Chaffee County,
on Mt. Antero and Mt. White, phenakite can be found in a
rarer prismatic (elongated) form. In both of these locations,
phenakite forms in pegmatites (very coarse granite formations). Phenakite is almost always colorless or white and it is
found with Colorado's other beryllium-bearing minerals, beryl
and betrandite. Beryl is the same hardness, but its forms are
always elongated and hexagonal (six-sided), differing from
phenakite's usual forms, and betrandite is softer.

WHERE TO LOOK: Mt. Antero and Mt. White, both in Chaffee
County, have pegmatite formations rich with phenakite, but
most of the area is now a nationally protected forest.

175

Crudely formed polybasite crystal clusters

Layered crystals

Siderite

Polybasite (metallic)

Pyrargyrite

Polybasite

ENVIRONMENT: Mine dumps

Occurrence

WHAT TO LOOK FOR: Small, silvery black, tabular
(flat, plate-like) crystals arranged into layered aggregates

SIZE: Polybasite specimens tend to be no larger than
your thumbnail

COLOR: Silvery black, rarely with dark red internal reflections

OCCURRENCE: Rare

NOTES: Polybasite is a rare ore of copper, silver and antimony
found in several mining districts throughout Colorado. Like
pyrargyrite, it is considered to be one of the "ruby silver"
minerals, which are ores of silver that have reddish colorations. While pyrargyrite is often bright translucent red,
polybasite is generally dark silvery black and only very fine
specimens will display deep, ruby-red flashes of color from
within the mineral. Crystals are rarely well formed, but appear
as flat, hexagonal (six-sided) disks. Most crystals are small and
crude, often intergrown as layered aggregates. Other
specimens will be rough, metallic masses formed atop other
silver-based minerals, such as pyrargyrite and even silver itself.
Identifying polybasite can be very difficult, as it often has a
very indistinct appearance and because it is the same color as
several other minerals. It can be very easily confused with
galena, pyrargyrite and silver, all of which have very similar
hardnesses. Galena, with which polybasite can be found, is
much heavier, much more abundant, and has a gray, rather
than black, streak. Pyrargyrite is generally more red in color,
but not always, so rely on streak color—pyrargyrite's is
reddish. Finally, masses of silver can look just like polybasite,
but silver will bend whereas polybasite will break.

WHERE TO LOOK: Try mine dumps in Clear Creek County.

Polycrase-(Y) crystal fragment

Brown surface coating

Crystal growth ridges

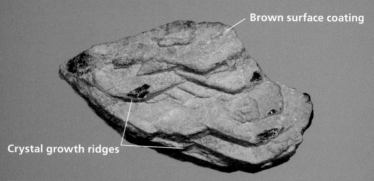

Massive polycrase-(Y)

Massive polycrase-(Y) fragments

Dull brown surface coating

Shiny black internal luster

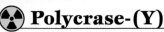

Polycrase-(Y)

HARDNESS: 5–6 **STREAK:** Yellowish to reddish brown

ENVIRONMENT: Mountains

Occurrence

WHAT TO LOOK FOR: Rough, crudely formed groupings of flat crystals embedded within pegmatite (a very coarse granite formation)

SIZE: Polycrase-(Y) crystals don't grow much longer than a few inches, though most specimens are thumbnail-sized

COLOR: Black to brown; often with pale brown surface coating

OCCURRENCE: Rare

NOTES: Polycrase-(Y) is one of the unique minerals that bears varying amounts of rare earth elements. Such minerals are called "REE minerals," for short. The REE element in the highest quantity is noted in parentheses after an REE mineral's name. In this case, yttrium is the dominant rare earth element in polycrase. As an REE mineral, polycrase-(Y) is a unique collectible, but good specimens are very rare. As with most REE minerals, it forms in pegmatites (very coarse granite formations), but mostly as black, brightly lustrous fragments or masses. Crystals are occasionally found, but they only appear as crude, rough wedges, often displaying angled growth ridges along their surface; these formed as the crystal was developing. In addition, crystals are often fragmented within the pegmatite, and extracting them results in broken specimens. While polycrase-(Y) is most often black to dark brown, it sometimes exhibits a light brown to yellow surface coating caused by weathering. Its luster is bright and nearly metallic, but it can be glassy as well. Fragments of other REE minerals, such as samarskite-(Y) can appear nearly identical, and telling each apart is next to impossible. Finally, polycrase-(Y) contains ample amounts of uranium, and it can be strongly radioactive as a result.

WHERE TO LOOK: Try looking in pegmatites in Chaffee County.

Purpurite coating

Purpurite coating

Purpurite

HARDNESS: 4–4.5 **STREAK:** Pale purple

ENVIRONMENT: Mountains, road cuts, mine dumps

Occurrence

WHAT TO LOOK FOR: Dark purple, dusty coatings on the surface of rocks or other minerals

SIZE: Purpurite forms as thin veins or coatings on rock, and specimens can measure several inches in size

COLOR: Purple to pink, brown

OCCURRENCE: Rare

NOTES: Purpurite, obviously named for its deep purple coloration, is a much-loved mineral among rock hounds because of its vivid color and attractive specimens. It forms in pegmatites (very coarse granite formations) as dark purple masses or veins in rock when pure, but it more commonly occurs as thin, dusty coatings that grow on the surface of other minerals or rocks. Many specimens are quite soft on the surface and will leave a purple dust on your hands. Purpurite forms after older minerals decay and weather; purpurite is a classic example of this type of formation. When lithiophilite, a rare lithium-bearing pegmatite mineral, decomposes, it lends its components to the creation of many other minerals. As the lithium leaches from the lithiophilite, its manganese and phosphorus content recombine to form purpurite. While you won't likely find lithiophilite in Colorado, purpurite is available to collectors and very easy to spot. And you won't easily confuse purpurite with other minerals, either. Few minerals have a similar deep, rich purple or pink color, and its hardness, streak, and the fact that it will easily leave your fingers purple are key identifying traits.

WHERE TO LOOK: Larimer County, in northern Colorado, contains several pegmatite outcrops where purpurite can be found.

Dark red pyrargyrite mass

Gray pyrargyrite mass

Lustrous red pyrargyrite coatings

Quartz

Pyrargyrite

HARDNESS: 2.5 **STREAK:** Purplish red

ENVIRONMENT: Mine dumps

Occurrence

WHAT TO LOOK FOR: Silvery red, soft, prismatic (elongated in one direction) crystals or veins with very high luster

SIZE: Pyrargyrite crystals occur thumbnail-sized and smaller while masses can be palm-sized and rarely larger

COLOR: Silver gray, red to dark red, reddish gray; darker when exposed to light

OCCURRENCE: Uncommon

NOTES: Pyrargyrite is one of the most common "ruby silvers," silver ores that exhibit a gem-like red, translucent coloration. When well formed, pyrargyrite develops elongated, prismatic crystals with broad, flat faces. However, like most minerals, pyrargyrite is rarely found in crystal form. Instead, you'll find cloudy red coatings or veins on quartz or other minerals in silver-mining districts. Its color can vary; many specimens are a dull gray with no distinct features. Still others are a dark reddish gray that appear mostly opaque until you shine a bright light on them. Nevertheless, in most specimens, pyrargyrite has a very bright luster and reflects light brightly, as if metallic. Of course, lesser-quality specimens are more dull, appearing glassy or greasy, but even these are often more lustrous than their surrounding minerals. Identifying pyrargyrite is fairly easy when good reddish color is present, but when specimens are dull, gray and mixed with other minerals, it can greatly resemble galena or polybasite. Luckily, pyrargyrite's dark reddish streak is distinctive and few other minerals share a similar streak color.

WHERE TO LOOK: Mine dumps in Park County are a good place to start.

Cubic pyrite crystals (brass yellow) with siderite crystals (brown)

Pyrite crystals

Embedded pyrite grains

Pyrite mass

Pyrite (brassy) on siderite (light brown)

Pyrite

HARDNESS: 6–6.5 **STREAK:** Greenish gray

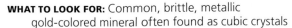

ENVIRONMENT: All environments

Occurrence

WHAT TO LOOK FOR: Common, brittle, metallic gold-colored mineral often found as cubic crystals

SIZE: Pyrite crystals are generally thumbnail-sized and smaller, but they can occasionally be found in much larger sizes

COLOR: Brass yellow to brownish yellow

OCCURRENCE: Very common

NOTES: Pyrite is one of the most common collectible minerals in the world. It is present in almost every geological environment and is as easy to identify as it is to find. Because of its great abundance, any mineral of brass-yellow color and metallic luster should be considered to be pyrite until you've determined otherwise. Pyrite crystals are very common and take the form of cubes, generally with striated (grooved) sides. These cubes will often be intergrown and have slightly curved faces. Massive or granular varieties are common as well and occur as large veins or masses embedded in rocks. Like marcasite, pyrite is a simple mixture of iron and sulfur, but even though both minerals have the same hardness, they are easy to distinguish, as marcasite does not form cubes and pyrite's streak has a greenish tint. Pyrite can also easily be visually confused with chalcopyrite, another extremely common brass-colored mineral, though chalcopyrite is softer. Pyrite has also been nicknamed "fool's gold" for decades by collectors, miners and tourists because it is thought to resemble gold. Gold, of course, is much rarer, more yellow in color, and is very soft. Pyrite often occurs with other iron-bearing minerals, including siderite and chalcopyrite, but also occurs with quartz, calcite and enargite.

WHERE TO LOOK: Literally any mine dump or rock outcrop in central Colorado will likely contain pyrite.

Radial crystal groupings

Radial crystals

Pyrophyllite mass (white)

Quartz

Pyrophyllite

HARDNESS: 1–2 **STREAK:** White

Occurrence

ENVIRONMENT: Mountains, mine dumps

WHAT TO LOOK FOR: Extremely soft, splintery aggregates of thin needles arranged into fan shapes

SIZE: Masses of pyrophyllite can be up to fist-sized and larger

COLOR: White to tan, gray, yellow to light green; orange to brown when impure

OCCURRENCE: Uncommon

NOTES: Pyrophyllite is an aluminum-based mineral closely related to talc, one of the softest minerals on earth. As such, pyrophyllite is extremely soft as well, and not only can you sink your fingernail deeply into its surface, merely overhandling a specimen will cause it to disintegrate. Crystals of pyrophyllite are virtually unknown. Instead, it forms as compact masses of soft grains or as fan-shaped aggregates of structures called folia. These folia are not crystals; rather they are thin, splintery leaf-like masses. They can be easily peeled off the rest of the specimen as they are highly flexible and loosely attached to each other. The characteristic radial "sprays" of pyrophyllite folia are highly distinctive, and combined with its extreme softness, you will never confuse pyrophyllite with any other mineral. It can form in several ways, but in Colorado, particularly in the San Juan Mountains, pyrophyllite forms as a result of other aluminum-rich rocks and minerals weathering and decomposing and can be found alongside quartz and clay minerals, such as montmorillonite. Pyrophyllite's colors range from white to tan or gray, but green and yellow varieties are common as well. Many specimens are orange or brown due to staining from iron minerals, particularly goethite.

WHERE TO LOOK: The San Juan Mountains in southwest Colorado have many deposits of pyrophyllite.

187

Actinolite coating

Diopside mass

Diopside mass

Augite (black) in diorite

Augite (black) in gabbro

Pyroxene group

HARDNESS: 5–6 **STREAK:** Greenish gray to white

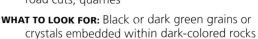

Occurrence

ENVIRONMENT: Mountains, plateaus, mine dumps, road cuts, quarries

WHAT TO LOOK FOR: Black or dark green grains or crystals embedded within dark-colored rocks

SIZE: Pyroxenes tend to form as small grains no larger than your thumbnail, but masses as large as your palm can be found

COLOR: Green, greenish gray to gray, brown to black

OCCURRENCE: Common

NOTES: The pyroxene group of minerals is closely related to the amphibole group. As such, pyroxenes play a large role in rock building, comprising much of the mineral content in dark rocks such as basalt and gabbro. Many pyroxenes, including augite, enstatite, aegirine and diopside are present in Colorado's rocks. Most of the pyroxenes generally appear as small, dark, glassy grains embedded in the rock. Augite, the most common pyroxene, is nearly always black and is most abundant in gabbro where it is visible as glassy, square crystals. Aegirine is similarly dark and glassy, but often appears as thin needles extending through the rock. Enstatite often lacks any kind of distinct shape when in rocks, but it is more brown in color. But of all Colorado pyroxenes, diopside is the most collectible and most easily identified. Common diopside is found in blocky masses with a dark, opaque green-gray color and is often found embedded in metamorphic rocks. "Chrome diopside" is a chromium-bearing variety that exhibits vivid emerald-green hues and is found as thumbnail-sized grains embedded in kimberlite, a rare variety of rock.

WHERE TO LOOK: Nearly any rock outcropping in central Colorado will give access to pyroxenes.

Quartz crystals

Needle quartz

Rough, massive quartz

Quartz

HARDNESS: 7 **STREAK:** White

ENVIRONMENT: All environments

Occurrence

WHAT TO LOOK FOR: Very abundant, light-colored, very hard crystals, masses, veins or pebbles

SIZE: Quartz ranges in size from masses the size of a basketball to crystal points smaller than a pea

COLOR: Colorless to white, gray, yellow to brown, purple, pink

OCCURRENCE: Very common

NOTES: Quartz is the single most abundant mineral on the planet, and it is absolutely essential for amateur and professional collectors alike to be able to identify quartz, recognize it in its various forms and understand its characteristics. Consisting of the elements silicon and oxygen (a combination called silica), quartz is generally colorless to white when very pure, but even small amounts of impurities can tint quartz any color. Well-formed quartz crystals are hexagonal (six-sided) and tipped with a point. Also known as "rock crystals," these glassy crystals are common and frequently found within cavities in rocks. Particularly thin and steep crystals are called needle quartz, a name that reflects their shape. Although quartz crystals are not rare, they are not as common as quartz's many other forms. Water-worn quartz pebbles are often found in rivers, while white quartz veins extend through rocks, and massive quartz fragments are strewn upon mountain roadsides. The most common quartz specimens are white or gray grains and masses in granite and other coarse-grained rocks. Identification of quartz is easy when you note its abundance, high hardness, six-sided crystal shape, and conchoidal fracture (when struck, circular cracks appear). It also produces a spark when chipped.

WHERE TO LOOK: Quartz can be found virtually anywhere in Colorado, but particularly in the central portions of the state.

Smoky quartz crystal points

Microcline feldspar

Rose quartz

Amethyst (purple)

Sphalerite (dark yellow)

Amethyst druse

Fluorite (green)

Quartz, varieties

HARDNESS: 7 **STREAK:** White

ENVIRONMENT: All environments

Occurrence

WHAT TO LOOK FOR: Very abundant, light-colored and very hard crystals, masses, veins or pebbles

SIZE: Quartz can be found in a wide range of sizes, from masses the size of a basketball to crystal points smaller than a pea

COLOR: White, gray to black, light to dark purple, pink

OCCURRENCE: Uncommon

NOTES: Quartz can occur in many colors; this varied coloration is caused by the inclusion of mineral impurities. Amethyst, found in Colorado's mountainous areas, is a purple variety owing its color to the interplay of aluminum, iron and natural irradiation. Rose quartz is a massive pink variety found in some of Colorado's pegmatite (very coarse granite) formations. But no variety of quartz is more famous in Colorado than smoky quartz, which appears gray to black. In the Pikes Peak area, smoky quartz is found intergrown with bright blue amazonite; this is arguably the most striking and beautiful occurrence of smoky quartz in the state. More commonly, however, it is found growing atop pink or tan microcline feldspar. It gets its color from aluminum impurities and natural irradiation. Quartz varies not only in color, but in formation as well. One form of quartz consists of crusts of thousands of tiny crystal points, often lining cavities within rocks or minerals. This type of formation is called druse, and while other minerals form druses, quartz forms druses more often than other minerals. Finally, agates, jasper and chalcedony are all common forms of microcrystalline quartz, masses that consist of quartz crystals too small to see.

WHERE TO LOOK: Mine dumps in the mountainous areas of central Colorado will give you the best shot at amethyst and smoky quartz. Agates are found along rivers in western Colorado.

Quartzite fragments

Quartzite containing green mica

Quartzite

HARDNESS: ~7 **STREAK:** N/A

Occurrence

ENVIRONMENT: Mountains, plateaus, mine dumps, quarries, riverbeds, road cuts

WHAT TO LOOK FOR: Very hard, light-colored, grainy-textured rock that shares many traits with quartz

SIZE: Quartzite forms massively and can be found in any size, from pebbles to boulders

COLOR: White to gray, yellow to brown; color varies with impurities

OCCURRENCE: Common

NOTES: When rocks are affected by heat and pressure (metamorphosis), they can change into completely different kinds of rocks. Limestone changes into marble, shale into slate, and sandstone turns into quartzite. Sandstone is a rock made up entirely of compacted sand, and sand consists primarily of quartz. So when sandstone is heated and compressed, it consolidates to form a tightly compacted mass of interlocking quartz grains. The resulting rock is called quartzite, a name that reflects its very high quartz content. Like quartz and other quartz-rich rocks like chert, quartzite is very hard and weather resistant. It has a slightly grainy appearance and is often white or gray in color, though it can be tinted other colors by impurities. Yellow or brown quartzite is colored by iron, and green quartzite, popular among collectors because of its ability to take a bright polish, can be colored both by serpentine inclusions as well as fuschite, a variety of green mica. Since quartzite can greatly resemble rough chert, it can be easy to misidentify. Look to the rock's surroundings—if it was found in a sedimentary area, it's likely chert. If the area appears to be more metamorphic or igneous in nature, it's probably quartzite.

WHERE TO LOOK: Try any metamorphic area in central Colorado.

Rhombohedral rhodochrosite crystal

Quartz

Loose rhombohedral crystal (⅝ inch across)

Rhodochrosite (pink) with chalcopyrite (metallic)

Massive rhodochrosite (pink) with galena (metallic gray)

Rhodochrosite

HARDNESS: 3.5–4 **STREAK:** White

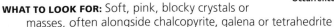

ENVIRONMENT: Mountains, mine dumps

Occurrence

WHAT TO LOOK FOR: Soft, pink, blocky crystals or masses, often alongside chalcopyrite, galena or tetrahedrite

SIZE: Masses or veins of rhodochrosite can be any size, but fine crystals are rarely larger than your thumbnail

COLOR: Rose pink to pale pink, red, reddish brown; darkens to brown on exposure to air and light

OCCURRENCE: Uncommon

NOTES: If you had to choose Colorado's single most famous mineral, it would have to be rhodochrosite. Many mines throughout the state have produced crystals of the beautiful pink mineral, but none are more famous than the Sweet Home Mine, in Alma. Specimens from this historic mine, regarded as the finest rhodochrosite locality in the world, cannot be obtained by amateurs, but they can be found in many shops for jaw-dropping prices. Sweet Home rhodochrosite is so unique because of its perfect rhombohedral crystals and its deep, rich, reddish pink color. Rhombohedrons, rhodochrosite's most common crystal form, are shapes resembling an angled, or "leaning," cube. This shape, combined with its pink color, make many specimens of rhodochrosite easy to identify. Poorly crystallized or massive specimens aren't much harder to identify—rhodochrosite's color will always be your first clue, and its low hardness will differentiate it from most other pink minerals. In addition, look for its commonly associated minerals, including quartz, chalcopyrite, galena, tetrahedrite and purple fluorite. In fact, rhodochrosite forms so frequently with these minerals that they can be found with virtually any specimen.

WHERE TO LOOK: Mine dumps in central Colorado's mountainous region are best, but fine specimens are extremely rare.

Massive rhodonite

Pyroxmangite

Pyroxmangite (pink) with galena (gray) and chalcopyrite (brass yellow)

Rhodonite/Pyroxmangite

HARDNESS: 5.5–6 **STREAK:** White

ENVIRONMENT: Mountains, mine dumps

Occurrence

WHAT TO LOOK FOR: Hard, pale pink masses, often intergrown with metallic minerals such as galena

SIZE: Masses of rhodonite can occur in any size, but they are generally smaller than an adult's fist

COLOR: Pale pink to deep pink, grayish pink, red to brownish red; often with a black dusty surface coating

OCCURRENCE: Uncommon

NOTES: Rhodonite is an easily identified manganese-bearing mineral found in mines throughout Colorado. Crystals of rhodonite are virtually unknown in the state, and it is instead found as rough masses, or chunks, often intergrown with many other minerals including galena, chalcopyrite and quartz. Rhodonite masses are opaque and generally very pale pink in color, though dusty black coatings of manganese oxides often coat specimens. It is generally formed in metamorphic rocks, so it is most frequently found in Colorado's mountainous mining districts. You're not likely to confuse rhodonite with anything else. Rose quartz may be a similar pink color, but is harder and more translucent, while rhodochrosite, another pink, manganese-bearing mineral is more translucent than rhodonite, much softer and is found crystallized more often. Some feldspars may be a similar pink, but they are generally more orange in color, more common, and slightly harder. Really the only mineral you'll confuse with rhodonite is pyroxmangite. Pyroxmangite is essentially identical to rhodonite, but forms under higher temperatures and pressures. With their identical appearances, hardnesses, and associated minerals, however, you won't be able to tell them apart.

WHERE TO LOOK: Most rhodonite comes from the San Juan Mountains.

Rough rhyolite

Pumice

Rhyolite

HARDNESS: 6–6.5 **STREAK:** N/A

Occurrence

ENVIRONMENT: Mountains, plateaus, mine dumps, road cuts, riverbeds

WHAT TO LOOK FOR: Fine-grained, brown or gray rock, often with many vesicles (gas bubbles) and colored bands or stripes

SIZE: Rhyolite can be found in any size, from pebbles to boulders

COLOR: Gray, reddish brown to brown, multicolored bands

OCCURRENCE: Common

NOTES: Like basalt, rhyolite is a fine-grained volcanic rock formed when lava (molten rock) spilled onto the earth's surface. The longer a rock has to cool and solidify, the larger the individual mineral grains within it become. On the earth's surface, rocks harden very quickly, which doesn't allow the minerals contained within the lava enough time to crystallize to a visible size. Therefore, the same body of lava has the potential to create different kinds of rocks. If the same lava that formed rhyolite had stayed within the earth and cooled very slowly, it would have formed granite. Rhyolite is fairly hard due to a large amount of quartz and feldspars in its composition. It also contains hornblende, a common amphibole, as well as various mica minerals. The result is a light-colored rock, ranging in color from light gray to brown, often with bands or stripes of color formed by the flowing motion of the lava. Unlike basalt, rhyolite's dark-colored cousin, rhyolite has visible (albeit tiny) mineral grains within it. Even though rhyolite cooled very rapidly, its lava is more viscous (thick or sticky) than basalt's, which helps it retain heat longer and therefore cool more slowly, forming larger mineral grains. Pumice is a glassy, frothy and bubbly variety of rhyolite, formed from gas-rich lava flows.

WHERE TO LOOK: Try looking near Ruby Mountain in Chaffee County.

Samarskite-(Y) fragment

Shiny black color

Grayish brown surface coating

Samarskite-(Y) (lustrous black)

Feldspar (orange)

☢ Samarskite-(Y)

HARDNESS: 5–6 **STREAK:** Dark reddish brown

ENVIRONMENT: Mountains

Occurrence

WHAT TO LOOK FOR: Dark, heavy, brightly lustrous radioactive masses embedded in pegmatites

SIZE: Most fragments of samarskite-(Y) are thumbnail-sized and smaller

COLOR: Black; often with light gray or brown surface coating

OCCURRENCE: Rare

NOTES: Samarskite-(Y) is one of Colorado's many minerals that contain rare earth elements (REE minerals, for short). REE minerals contain some of the earth's rarest elements, such as cerium, ytterbium and lanthanum. These elements can exist in varying amounts within the mineral, so the REE element in the highest quantity is noted in parentheses after an REE mineral's name. Samarskite-(Y) contains yttrium as the dominant rare earth element, but also contains niobium, tantalum and varying amounts of thorium and uranium, making many specimens of samarskite-(Y) highly radioactive. Samarskite-(Y) forms exclusively in pegmatites (very coarse granite formations) and is often embedded between masses of quartz and feldspars. Crystals are extremely rare and virtually always appears as black, brightly lustrous, nearly metallic masses, often coated with a thin, grayish brown film on the outer surface that you have to break or scratch through in order to see the mineral's true color. These masses are often quite heavy for their size and, as mentioned above, many are radioactive. The problem with identification is that several other REE minerals—particularly polycrase-(Y)—form in almost identical masses with similar hardness and streak. Unfortunately, nothing short of laboratory analysis will positively identify samarskite-(Y).

WHERE TO LOOK: Try pegmatite formations in Fremont County.

Rough sandstone fragments

Banded sandstone (cut flat)

Sandstone

HARDNESS: N/A **STREAK:** N/A

ENVIRONMENT: Plateaus, fields, road cuts, mine dumps, quarries

Occurrence

WHAT TO LOOK FOR: Layered rocks that appear to be made of sand and have a gritty, rough feel

SIZE: Sandstone can occur in any size, from pebbles to boulders

COLOR: White to gray, yellow to brown, red to reddish brown

OCCURRENCE: Very common

NOTES: Sandstone is exactly what it sounds like—rock made of sand. It formed in sedimentary areas where a body of water or wind action caused grains of sand to build up in large amounts. Later, finely grained sediment settled in between the grains and began to cement the sand together. The result of this process is a coarse, grainy rock with a gritty, rough feel. You can also often separate individual grains from the rock, depending on how well the specimen has been cemented. The color of sandstone depends on two factors: the sand itself and the type of cement binding the grains together. The cement is primarily responsible for sandstone's color and often contains iron impurities that turn it yellow, brown, orange or red. Despite the fact that sand consists primarily of light-colored quartz, many pink, yellow or brown feldspar grains are also present, which can affect sandstone's color as well. If the sandstone contains a large amount of feldspar grains in addition to the quartz grains, it is called arkose. Both sandstone and arkose are found in flat, sedimentary areas, particularly in the northwestern corner of Colorado. Sandstone often exhibits bands or stripes of varying color caused by the layering of new sand layers over old ones. And as a sedimentary rock, sandstone can also contain fossils.

WHERE TO LOOK: Look for sandstone in northwestern and eastern Colorado.

Scheelite crystals (brown) on fluorite

Loose crystal

Above specimen under short-wave ultraviolet light

Scheelite

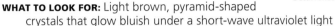

HARDNESS: 4.5–5 **STREAK:** White to pale yellow

ENVIRONMENT: Mountains, mine dumps

Occurrence

WHAT TO LOOK FOR: Light brown, pyramid-shaped crystals that glow bluish under a short-wave ultraviolet light

SIZE: Masses or veins of scheelite can be almost any size while crystals remain smaller than your thumbnail

COLOR: Brown, white to gray, yellow to orange

OCCURRENCE: Uncommon

NOTES: Scheelite, an uncommon calcium- and tungsten-bearing mineral, is a favorite among rock hounds for its well-formed crystals and unique fluorescent habit. Brown, gray or orange scheelite has been found in many mines across Colorado, and finely formed, steeply pointed pyramid-shaped crystals can be found in most scheelite localities. But even if a specimen is found as a massive chunk rather than well-developed crystals, identification is easy because of scheelite's trademark fluorescence. When a short-wave ultraviolet light is shone on scheelite, the specimen will glow brightly in shades of bluish white. Some specimens may glow a creamy yellow color if some amount of the element molybdenum is present. In normal light, most Colorado scheelite is brown to gray and translucent, though few specimens are flawless and crystals are usually cloudy and internally fractured. Crystal points of scheelite can be found growing on fluorite or quartz, though confusing scheelite with those minerals is unlikely. Quartz is harder, and even though fluorite can be fluorescent as well, it is softer than scheelite. Scheelite's color and crystal shape also resemble zircon's, but zircon is much harder and forms in granite, but scheelite does not.

WHERE TO LOOK: Mine dumps in Boulder and Ouray counties have produced well-formed crystalline specimens.

Antigorite (green)

Chrysotile (fibrous)

Serpentinite

Serpentine group

HARDNESS: 2.5–5 **STREAK:** White

ENVIRONMENT: Mountains, plateaus, mine dumps, road cuts, quarries

WHAT TO LOOK FOR: Greenish or yellowish minerals with a distinctly "greasy" feel, sometimes with flaky, fibrous crystals

SIZE: Masses of serpentine can be quite large, but specimens are generally fist-sized and smaller

COLOR: Yellow to green, golden yellow, olive green

OCCURRENCE: Common

NOTES: The serpentines are a common group of green, magnesium-bearing minerals. Two of Colorado's serpentines, antigorite and lizardite, are green to yellow minerals that form in solid, compact masses and feel "greasy" to the touch. Antigorite forms in smooth, fairly hard masses that are generally translucent with patches of cloudy coloration. Lizardite is softer, more opaque, and exhibits a scaly, flaky surface structure. Telling the two apart is not particularly easy unless a definite hardness can be measured. Antigorite tends to be in the upper end of the range—3.5 to 5—while lizardite remains softer than 3.5. A third variety of serpentine, called chrysotile, is also common. Chrysotile serpentine is fibrous, silky mineral forming veins of tiny, thin, needle-like crystals. These flexible, easily separated fibers are a type of mineral formation called asbestos. Asbestos fibers can be so thin and small that they cannot be easily seen. They can also become airborne and present a cancer risk if inhaled. Always wear a mask or respirator when working with chrysotile serpentine. Serpentinite is a dark, green, serpentine-rich rock exhibiting serpentine's trademark "greasy" feel.

WHERE TO LOOK: Try Colorado's metamorphic mountain regions.

209

Oil shale

Rough shale

Slate

Shale

Shale/Slate

HARDNESS: N/A **STREAK:** N/A

ENVIRONMENT: Plateaus, fields, road cuts, quarries

WHAT TO LOOK FOR: Fine-grained, layered rocks that occur in large sheets and can be split apart along their layers

SIZE: Shale and slate both occur in large sheets, sometimes miles across in size

COLOR: Yellowish brown to brown, tan, gray to black

OCCURRENCE: Shale is very common; slate is uncommon

NOTES: Shale is a sedimentary rock made of nearly microscopic, silt-sized grains of weathered, broken-down minerals, unlike sandstone which is formed from comparatively large sand-sized sediments. The various individual minerals in shale are invisible except under very high magnification; these are primarily clay minerals such as kaolinite and montmorillonite, as well as some tiny grains of quartz and micas. These incredibly small pieces of sediment are easily carried by even gentle water currents, so very still or deep water is required for silt to settle to the bottom, where it later can be compressed and consolidated into rock. Shale beds settle and compact over time; this creates layers in the shale that are easy to split apart into thin, flat sheets. Because of its origins as sediment in water, shale can also contain fossils, such as leaves and snail shells. Some shale contains bitumen, a naturally occurring tar-like material that can be distilled to yield oil. Appropriately, this variety is called "oil shale" and is dark gray to black. When shales are heated and further compressed, they turn into slate, a dark gray, brittle rock famous for its use in chalkboards. Slate exhibits layers as well, and although the layers are much thinner and more tightly compact, they can still be split apart.

WHERE TO LOOK: Shale can be found in low-lying, sedimentary areas of the state, such as the entire eastern side of Colorado.

Occurrence

211

Siderite crystal aggregate

Siderite (tan) with polybasite (gray)

Siderite crystal aggregate with pyrite (brassy yellow)

Siderite

HARDNESS: 3.5–4 **STREAK:** White to pale yellow

ENVIRONMENT: Mountains, plateaus, plains, mine dumps, quarries, road cuts

WHAT TO LOOK FOR: Light brown, blocky masses or small, rounded, disk-shaped crystals alongside pyrite

SIZE: Masses can be up to fist-sized or larger while crystals remain small, rarely larger than a pea

COLOR: Light to dark brown, cream-colored, white

OCCURRENCE: Common

Occurrence

NOTES: Siderite is a common mineral very closely related to calcite, dolomite and rhodochrosite. While its mineral cousins may be more attractive and collectible, siderite is a diverse and important mineral. When well crystallized, this iron-rich mineral forms small, curved blades or disk-shaped crystals with pearly luster, often intergrown in complex aggregates. However, as with most minerals, crystals are rarer than massive specimens, which can be quite large in size. Rough, irregular siderite growths are quite common and can often be the base from which other minerals grow; this is true for other iron minerals such as pyrite and chalcopyrite. Massive siderite can also greatly resemble other rocks or minerals, such as calcite, alunite and limestone, though there are a few tests that will help. Siderite will only slightly effervesce (fizz) in vinegar while calcite and limestone will do so vigorously as they dissolve in the acid. Alunite will not effervesce at all. Magnetism is another trait you can test. Siderite becomes weakly magnetic after it has been heated over a flame.

WHERE TO LOOK: Siderite masses and crystals can be found in sedimentary areas, particularly within shale, as well as in ore districts. In particular, siderite is abundant in Eagle County and can be found in virtually any mine dump.

Silver nugget (partially coated by dark gray acanthite tarnish)

Slice of silver ore

Silver wire

Silver flake in limonite

Silver

HARDNESS: 2.5–3 **STREAK:** Silver gray to white

ENVIRONMENT: Mountains, plateaus, mine dumps, riverbeds

WHAT TO LOOK FOR: Soft, bendable, white silvery metal, most often with dark gray surface tarnish

SIZE: Specimens of silver are rarely larger than your thumbnail; silver ore, or silver-bearing rock, can occur in any size

COLOR: Silver-white or gray; dark gray or black surface tarnish when exposed to air

OCCURRENCE: Uncommon

NOTES: Silver, the white metallic element famous for its value and use in jewelry, is widespread throughout Colorado and over one hundred years of mining has produced nearly a billion dollar's worth of the metal. Most silver is mined in the form of ore, especially in the Leadville area where much of the state's silver wealth has been extracted from rocks containing silver-rich galena, chalcopyrite and quartz. Pure, or native, silver can also be found as nuggets, flakes or crystals embedded in rocks or other minerals, especially limonite and quartz. Silver wire is perhaps the most interesting variety of silver. These bizarre, twisting crystals are actually very deformed and elongated cubes of silver. Whatever the form a silver specimen takes, it is always easy to identify. Silver is always highly malleable, meaning it will bend and change shape instead of being brittle and breaking under stress. Its trademark color is also distinctive, but is often tarnished to a dark gray or brownish coloration. This surface tarnish is actually a mineral called acanthite, and scratching through it will reveal silver's true color.

WHERE TO LOOK: Pitkin County and Clear Creek County produce ample amounts of silver, but the mine dumps in the Creede area of Mineral County are known for their fine specimens.

Smithsonite

Quartz

Smithsonite

Smithsonite on quartz

Smithsonite "blobs" (green) on quartz

Smithsonite

HARDNESS: 4–4.5 **STREAK:** White

ENVIRONMENT: Mountains, mine dumps

WHAT TO LOOK FOR: Gray to green botryoidal (grape-like) coatings or masses growing on quartz or cerussite

SIZE: Masses of smithsonite can be a wide range of sizes, from thumbnail- to palm-sized

COLOR: Green to dark green, blue to blue-green, gray

OCCURRENCE: Uncommon

NOTES: Smithsonite, named for James Smithson, founder of the Smithsonian Institution, is an ore of zinc that is closely related to calcite, siderite and rhodochrosite. Crystals of smithsonite are exceedingly rare and it is instead found as thin, botryoidal (grape-like) coatings or masses growing on top of other minerals, like quartz and cerussite. Deep in several of Colorado's mines, smithsonite was found as large encrustations on cave walls, even forming stalactites (hanging, icicle-like formations). These rich deposits proved to be a valuable source of zinc. Specimens can be found in a range of colors, but the green or blue varieties are most desirable to collectors. Many "blobs" or coatings of smithsonite are so pale in color that they can easily be overlooked. When you think you've found a smithsonite specimen, determining its identity is fairly easy. If a sample has smithsonite's characteristic bluish green color, the only similarly colored minerals are chrysocolla, turquoise and occasionally hemimorphite. Chrysocolla is softer than smithsonite, turquoise is harder, and both are opaque, vividly colored copper minerals. Hemimorphite is generally gray and only rarely green or blue. Hemimorphite is also slightly harder, commonly crystallized, and doesn't often form botryoidal masses.

WHERE TO LOOK: The Leadville area produces many smithsonite specimens, and they can still be found in mine dumps.

217

Sphalerite (black)

Mass of sphalerite crystals

Chalcopyrite (brassy brown)

Close-up of crystal mass

Well-crystallized sphalerite

Triangular crystal face

Chalcopyrite (brass yellow)

Sphalerite

HARDNESS: 3.5-4   **STREAK:** Light brown

ENVIRONMENT: Mountains, plateaus, mine dumps

WHAT TO LOOK FOR: Dark, glassy triangular crystals or masses occurring alongside chalcopyrite and galena

SIZE: Sphalerite crystals generally measure no larger than your thumbnail while masses can be any size

COLOR: Gray to black, dark green to greenish yellow, yellow to brown, dark red

OCCURRENCE: Common

NOTES: Sphalerite, the most important ore of zinc, is a common mineral that every rock hound should be able to identify. It forms in a number of environments, such as in limestone formations, but sphalerite is particularly common in ore veins where it is very frequently intergrown with galena, chalcopyrite and pyrite. Crystals are abundant and can be found as pyramid-shaped points or wedges with triangular faces. Rough, poorly formed masses are common as well, so you must learn how to recognize sphalerite's telltale characteristics. Sphalerite is generally always a dark color, appearing almost black and metallic on first glance. However, on closer inspection, especially under a bright light, you will notice that nearly all specimens are actually a translucent dark green or yellow color, or, more rarely, red. Thinner or broken fragments are likely to show the specimen's true color more easily. Specimens always have a high luster and brightly reflect light unless dulled by other mineral impurities. As mentioned above, sphalerite is very commonly intergrown with chalcopyrite and galena, and any dark, glassy crystals found along with those minerals should be assumed to be sphalerite until you know otherwise.

WHERE TO LOOK: Try mine dumps in ore mining areas, especially Lake, Mineral and Ouray counties.

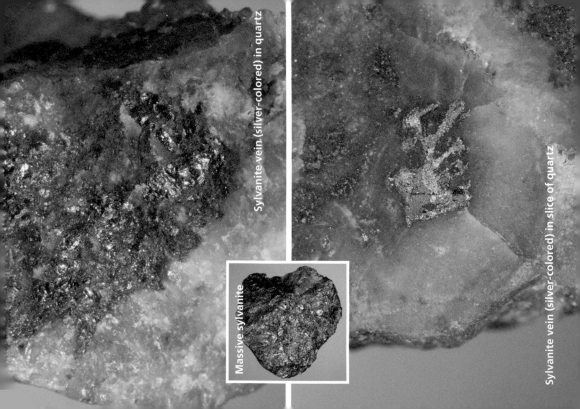

Sylvanite vein (silver-colored) in quartz

Sylvanite vein (silver-colored) in slice of quartz

Massive sylvanite

Sylvanite

HARDNESS: 1.5–2   **STREAK:** Silver-gray

ENVIRONMENT: Mountains, mine dumps

WHAT TO LOOK FOR: Very soft, bright silvery metallic masses or veins embedded within rock

Occurrence

SIZE: Sylvanite specimens are rarely larger than your thumbnail and most individual crystals or veins are very small

COLOR: Silver-gray, silver-white, silver-yellow

OCCURRENCE: Very rare

NOTES: Sylvanite is one of Colorado's legendary telluride minerals, which are combinations of rare elements and tellurium. Sylvanite contains both gold and silver, making it a very valuable ore that was widely mined throughout the state for decades. Along with calaverite, another gold-bearing telluride mineral, sylvanite has been one of the most economically important minerals in all of Colorado, producing millions of dollars in gold. It is found primarily in both Teller and Boulder Counties, though most mines that produce sylvanite are closed or privately owned. But if you do think that you've found a piece of sylvanite, determining its identity is fairly easy thanks to its extremely low hardness. The silvery white metallic mineral greatly resembles its telluride cousin, calaverite, as well as arsenopyrite, but sylvanite is much softer than both minerals. Sylvanite crystals are extremely rare, and most specimens consist of veins running through quartz or fluorite. Its tendency to form in veins combined with its low hardness could possibly be confused with molybdenite, but this is unlikely. Molybdenite has a dark bluish gray appearance and is not found in the same types of rock. When in doubt, however, a streak test will certainly help.

WHERE TO LOOK: Mines in Boulder and Teller counties produced more sylvanite than anywhere else in the state.

Tetrahedrite crystal (3mm) on dolomite (brown)

Triangular crystal face

Pyrite crystals (brassy yellow)

Tetrahedrite (metallic black)

Tetrahedrite group

HARDNESS: 3.5-4 **STREAK:** Black to dark brown

ENVIRONMENT: Mountains, plateaus, mine dumps

WHAT TO LOOK FOR: Dark-colored, sharp triangular crystals or masses alongside sphalerite, dolomite or quartz

SIZE: Tetrahedrite crystals rarely grow larger than a pea, but masses can be thumbnail-sized and rarely larger

COLOR: Steel-gray to black, silver-gray

OCCURRENCE: Uncommon

NOTES: The tetrahedrite group consists primarily of three minerals, tennantite, freibergite and tetrahedrite, all of which can be found in Colorado. The group gets its name from the shape of their tetrahedral crystals, forming as four-faced, triangular pyramids. Tennantite and freibergite are much rarer than tetrahedrite, but since the group members are virtually indistinguishable from each other, you'll almost never know with certainty which mineral is which. Due to their identical crystal shapes, hardnesses and similar streak colors, only laboratory testing will enable you to differentiate them, so nearly all specimens are simply labelled "tetrahedrite." While it may not be exactly correct for all specimens, it is an acceptable solution considering how similar the minerals are and how much more common tetrahedrite is than the others. Tetrahedrite's metallic gray crystals are often very small, rarely growing larger than a half inch, which can aid in identification. You could, however, mistake a crystal of sphalerite for tetrahedrite due to their similar crystal shapes, identical hardness, and frequent occurrence with the same associated minerals, such as pyrite and quartz. Sphalerite, however, is translucent and glassy, rather than metallic.

WHERE TO LOOK: Mine dumps in Hinsdale and Park counties should yield tetrahedrite minerals.

Occurrence

223

Thorite (brick red) grains in granite

Thorite (brick red) in granite

Waxy surface luster

Thorite

HARDNESS: 4.5 **STREAK:** Light brown

ENVIRONMENT: Mountains

WHAT TO LOOK FOR: Reddish brown radioactive masses within granite or pegmatites (very coarse granite formations)

SIZE: Masses of thorite are rarely larger than a golf ball; most specimens are grains smaller than your thumbnail

COLOR: Brick-red, reddish brown to dark brown

OCCURRENCE: Rare

NOTES: Colorado is home to many radioactive minerals, but virtually all of them derive their dangerous, ionizing energy from one of two elements: uranium or thorium. Thorium, the rarer of the two, is present in far fewer minerals than uranium, but it isn't any less dangerous. Thorite is a mixture of thorium, silicon and oxygen and is the most common thorium-bearing mineral. Its blocky, square crystals are very rare in Colorado and instead thorite commonly forms as brick-red to brown masses or grains embedded within granite and pegmatite (very coarse granite) formations, as well as some metamorphic rocks. Thorite masses often have a waxy surface appearance and texture, not unlike that of a jasper. You won't confuse the two, however, because jasper doesn't form in the same kinds of rock. In fact, thorite's occurrence in granites or pegmatites, along with its color and strong radioactivity are enough to identify most specimens. Pale orange or yellowish brown specimens of thorite can also resemble feldspars, as both minerals are often found embedded in rocks. Nevertheless, feldspars are never radioactive. As with all radioactive minerals, take proper care when collecting and storing thorite.

WHERE TO LOOK: Try the mountain ranges in Custer and Fremont counties, southwest of Pueblo.

Occurrence

Titanite crystal fragments from a pegmatite formation

Titanite crystal

Titanite crystal

Uralite

Titanite (Sphene)

HARDNESS: 5–5.5 **STREAK:** White

Occurrence

ENVIRONMENT: Mountains, plateaus, mine dumps, riverbeds

WHAT TO LOOK FOR: Very small, brown or yellow wedge-shaped crystals or grains embedded within rock

SIZE: Both crystals and embedded grains of titanite are small and are rarely found larger than your thumbnail

COLOR: Gray, yellow to yellowish brown, reddish brown, brown

OCCURRENCE: Uncommon

NOTES: Many older rock hounds may be more familiar with titanite's former name, sphene. Its name was changed in 1982 to better reflect its titanium-rich composition. Today, the name sphene is still often used, especially when referring to a cut titanite gemstone used in jewelry. Long favored by collectors for its brightly lustrous, sharply pointed wedge-shaped crystals, titanite is a fairly common collectible in Colorado, but primarily occurs as tiny grains embedded in metamorphic rocks. Small, pea-sized titanite crystals can be found in cavities, often growing on top of other minerals, such as epidote and amphiboles. One of its most interesting occurrences are small, yellowish titanite crystal points that grow in between crystals of uralite, a grayish green variety of amphibole. But the only crystals of any size are those found in pegmatites (very coarse granite formations), where reddish brown masses and crude crystals can grow to several inches in length. Titanite grains or masses within rocks can be hard to notice, let alone identify, but crystals are easy to spot. The only mineral you'll likely confuse with a titanite crystal is sphalerite, but sphalerite is softer.

WHERE TO LOOK: Pegmatites in Eagle County produce larger crystals, while mountainous areas in Chaffee County are known for small crystals amid uralite.

Topaz crystal

Rhyolite

Topaz crystal on matrix

Rough, massive topaz fragments

Topaz

HARDNESS: 8   **STREAK:** White

ENVIRONMENT: Mountains, plateaus, mine dumps, road cuts, riverbeds

WHAT TO LOOK FOR: Very hard crystals or masses found within igneous rocks, particularly pegmatites (very coarse granite formations) and rhyolite

SIZE: Most topaz specimens are crystals smaller than your thumbnail, but masses larger than an adult's fist have been found; very rarely, crystals measuring several inches turn up

COLOR: Colorless to white, bluish gray to gray, yellow to brown

OCCURRENCE: Uncommon

NOTES: Topaz, famous for being November's birthstone, is one of Colorado's most collectible and desired minerals. It is commonly found as well-formed, glassy and translucent crystals that are easy to identify. Its elongated, prismatic crystals are tipped with a pyramid-shaped point or wedge and form best in cavities within volcanic rocks, particularly rhyolite. Ruby Mountain, near Nathrop, is a particularly well-known topaz locality, and excellent specimens are frequently found alongside garnets in rhyolite. Specimens of topaz are regularly colorless, gray or brown in color, but blue and white varieties are found throughout Colorado. Topaz also forms in pegmatites (very coarse granite formations) as large irregular masses and within granite as rough, crude crystals that generally don't come close to the quality of those formed in rhyolite. Massive or poorly crystallized specimens can resemble quartz or beryl, but topaz is much harder than quartz and just slightly harder than beryl. Also, be aware that brown topaz exposed to sunlight often fades to colorless.

WHERE TO LOOK: Ruby Mountain, in Chaffee County, and Pikes Peak are well known for fine crystals.

229

Pegmatite with embedded tourmaline

Quartz (white)

Schorl (black)

Muscovite mica

Schorl embedded in quartz

Elbaite (dark green) in quartz

Dravite

Elbaite

Tourmaline group

HARDNESS: 7–7.5 **STREAK:** White

Occurrence

ENVIRONMENT: Mountains, plateaus, mine dumps, road cuts

WHAT TO LOOK FOR: Black, long, slender and striated (grooved) crystals and masses embedded in metamorphic rocks and pegmatites (very coarse granite formations)

SIZE: Tourmaline crystals are generally small, no more than half an inch in diameter and less than six inches long, at their largest

COLOR: Black, dark blue to green, dark brown, rarely white

OCCURRENCE: Common

NOTES: The tourmaline group consists of over a dozen different minerals formed by extremely complex chemical compositions including aluminum, boron and silicon. In Colorado, three tourmaline minerals are known to occur: schorl, dravite and elbaite, all of which are very hard. Of the three, schorl and elbaite are the easiest tourmalines to identify. Schorl is always black and forms elongated crystals, sometimes several inches in length while still remaining very thin. The surfaces of these crystals are deeply striated (grooved); looking at a crystal's cross-section will reveal that the crystals grow in a triangular shape. Elbaite shares most of the same traits but can be many different colors. In Colorado, dark green, blue, and even rare white elbaite crystals are known. Schorl can form as embedded crystals in granite or pegmatites (very coarse granite formations), while elbaite is generally only found in pegmatites, particularly with quartz. Dravite can be black to brown and doesn't share the same long, slender crystal shape. Instead, it forms stubby, short crystals with a hexagonal cross-section.

WHERE TO LOOK: Try mine dumps in Chaffee County for dravite, and Clear Creek County pegmatites for elbaite.

Tuff

Fragments of volcanic glass (black)

Welded tuff

Tuff

HARDNESS: <5 **STREAK:** N/A

ENVIRONMENT: Plateaus, fields, road cuts, quarries

Occurrence

WHAT TO LOOK FOR: Light-colored, gritty rock, often exhibiting layers and fragments of black glass within

SIZE: Tuff is a rock that can be found in any size, from pebbles to mountains

COLOR: White to gray, yellow to brown, often with spots of black

OCCURRENCE: Common

NOTES: Tuff is an igneous rock formed when an eruption threw volcanic ash and small fragments of pulverized rock over the landscape. Different forms of tuff can exist depending on the conditions of its formation. In many volcanic eruptions, ash simply fell from the air and settled into thick beds, compacting and consolidating over time to form a soft, gritty-feeling rock. Other tuffs are formed from much hotter eruptions. When the ash particles from these eruptions fell to the ground, they were still very hot and partially molten, causing them to adhere to each other on contact. The result is called welded tuff, and it is much harder, more compact, and more firmly held together. Both types of tuff are quite porous, which can allow mineral solutions to seep into the rock. Quartz often gets into these porous spaces which makes the tuff much harder in a process called induration. Over time, several eruptions and ash falls may have occurred, resulting in a variety of layers (and tuff colors) throughout a tuff formation. But because of all of these variabilities in hardness and color, tuff can be difficult to identify. Tuffs are all very fine grained, but in most speci- mens, tiny, dark, glassy fragments of obsidian (volcanic glass) are also present. This will be one of your biggest clues.

WHERE TO LOOK: Look in western Colorado's plateau region, where volcanic activity took place long ago.

Turquoise fragments

Turquoise coating on rock

Turquoise vein in quartz

Specimens courtesy of Irene Witmer

Turquoise coating on the surfaces of rock

Turquoise

HARDNESS: 5–6 **STREAK:** Pale green

ENVIRONMENT: Mountains, mine dumps

Occurrence

WHAT TO LOOK FOR: Masses of hard, blue material embedded within rock as grains or veins

SIZE: Masses of turquoise occur smaller than an adult's palm

COLOR: Light blue to dark blue, greenish blue

OCCURRENCE: Uncommon

NOTES: Of all the many copper-bearing minerals, turquoise is perhaps the most well known due to its longtime use in jewelry. However, its use as a decorative stone isn't a contemporary idea; turquoise has been mined, carved and worn for thousands of years throughout the world. As with many other blue or green minerals, turquoise gets its color from its copper content, but differing amounts of copper content or impurities can cause a wide range of colors to appear. In fact, the color of turquoise can change drastically between localities, and many serious turquoise collectors can tell where a specimen originated just by looking at its color. When hunting for turquoise, its color will be your biggest clue, though other copper-bearing minerals can exhibit similar color as well, such as chrysocolla and malachite. Luckily, turquoise is much harder than all similarly colored minerals, and the way in which it forms removes any doubt. Turquoise crystals are non-existent in Colorado and instead it forms as masses filling pockets or veins in rock. Since it forms as a result of other minerals weathering in acidic conditions, it can be found in or on many kinds of rock, including granite. Pale blue specimens are considered to be less desirable than the deep, dark, bluish green specimens.

WHERE TO LOOK: Lake and Teller counties have produced a lot of turquoise, some lying loose on the surface.

fragment of massive uraninite

Uraninite (black)
with calcite (white)

Tree growth layers

Uraninite replacing wood

☢ Uraninite

ENVIRONMENT: Mountains, plateaus, mine dumps

WHAT TO LOOK FOR: Black masses of sooty, dusty radioactive material embedded in rock

SIZE: Uraninite most commonly forms massively and can be found in any size, from tiny grains to large masses measuring several feet in size

COLOR: Black to gray, dark brown

OCCURRENCE: Common

NOTES: Uraninite consists of over eighty percent uranium, making it one of Colorado's most strongly radioactive minerals. It is also Colorado's most common radioactive mineral and can be found in a range of different mineral deposits. In fact, if your Geiger counter is detecting radioactivity, you should assume uraninite is present before any other mineral. Crystals of uraninite are very rare, but still can be found in pegmatites (very coarse granite formations) as small, black cubes. Irregularly shaped masses, veins or grains of uraninite embedded in rocks are far more common. Most uraninite specimens won't exhibit any obvious mineral structure, making many specimens easy to overlook. For this reason, a Geiger counter is essential both for identifying uraninite as well as for your safety. Uraninite is generally always black, but often has a brownish tint. It also is normally dull in appearance, but a very fresh or pure specimen may have a metallic luster. Coffinite, a similar radioactive black mineral, is rarer, has a black streak, and is generally always brightly lustrous. One of uraninite's most interesting habits is when it replaces fossil wood, which makes for black, heavy, radioactive petrified wood.

WHERE TO LOOK: Boulder and Clear Creek counties have produced huge amounts of uraninite, particularly from ore veins.

Occurrence

237

Uranophane crystal clusters (yellow)

Close-up of radiating crystal aggregate (3mm)

Uranophane coating (yellow) on rock

Uranophane

HARDNESS: 2–3 **STREAK:** Pale yellow

ENVIRONMENT: Plateaus, mountains, mine dumps

Occurrence

WHAT TO LOOK FOR: Tiny, bright yellow, needle-like crystal aggregates or crusts

SIZE: Individual crystals are very small, measuring shorter than a millimeter in most cases, but crystal aggregates or urano-phane crusts can be thumbnail-sized and rarely larger

COLOR: Light to dark yellow, lemon yellow, greenish yellow, pale orange

OCCURRENCE: Rare

NOTES: One of Colorado's most attractive radioactive minerals is uranophane, which commonly forms tiny needle-like crystals arranged into radiating or fan-shaped aggregates growing on the surface of rocks. These minute, elongated needles rarely reach a quarter-inch in length and are always vivid shades of yellow or orange. Even as a well-formed crystal grouping, however, most specimens are very small, requiring sharp eyes to spot. When handling uranophane specimens, always wear thick leather gloves. The thin, needle-like crystals have been known to act as radioactive slivers, easily embedding in your skin. Occasionally, you may find a massive coating or a yellow stain on a rock that may be uranophane, but distinguishing a poorly formed or massive specimen of uranophane from similar radioactive yellow minerals such as carnotite, meta-autunite and metatyuyamunite, can be nearly impossible outside of a laboratory. This is because such minerals all share similar colors, hardnesses and streaks. If there are no clues as to a massive specimen's identity, assume it is either meta-autunite or carnotite, as they are more common.

WHERE TO LOOK: Try mine dumps in the Uravan Belt in southwest-ern Colorado, particularly in western San Miguel and Mon-trose counties, but be sure to bring your Geiger counter.

Cluster of vesuvianite crystals

Individual vesuvianite crystal

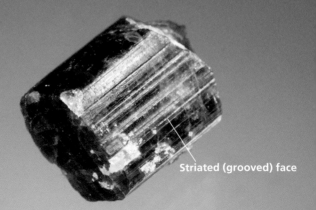

Striated (grooved) face

Vesuvianite (Idocrase)

HARDNESS: 6.5 **STREAK:** White

ENVIRONMENT: Mountains

Occurrence

WHAT TO LOOK FOR: Brownish, glassy, transparent crystals found within metamorphic rocks

SIZE: Vesuvianite crystals are commonly thumbnail-sized, but rarely larger, many specimens are small pea-sized grains or masses

COLOR: Brown to yellow, light green to blue, white to colorless

OCCURRENCE: Rare

NOTES: Commonly referred to as idocrase when used as a gemstone, vesuvianite is named for Mount Vesuvius, the Italian volcano infamous for burying the ancient city of Pompeii. This rare mineral can be found in some of Colorado's calcium-rich metamorphic rocks both as small, poorly formed grains and perfectly developed, thumbnail-sized crystals. When crystallized, vesuvianite is fairly easy to identify. Most crystals are stubby and rectangular, exhibiting striated (grooved) faces and a square cross-section. Many of these crystals are flat-topped, while others taper off to a blunt point. Crystal clusters and groupings are common, as well. Its color generally ranges from brown to dark golden yellow, but blue, green, and even colorless crystals have been found, albeit rarely. Grains or poorly formed crystals could be confused with several other minerals, particularly zircon, dravite, and possibly some garnets. Zircons are considerably harder and tend to form in granite, whereas vesuvianite does not. Dravite, a member of the tourmaline group, is harder as well. Vesuvianite can also occur alongside garnets, but garnets tend to be more red in color, form round, ball-like crystals, and are generally slightly harder.

WHERE TO LOOK: Metamorphic rocks in Gunnison County have produced some of the best vesuvianites in the US.

Volborthite (bright green)

Malachite (blue-green)

Volborthite and malachite in sandstone

Volborthite (bright green) in sandstone (white)

Volborthite

HARDNESS: 3.5 **STREAK:** Light green

ENVIRONMENT: Plateaus, mine dumps

Occurrence

WHAT TO LOOK FOR: Thin, green plate-like crystals or crusts, particularly within sandstone

SIZE: Individual volborthite crystals are rarely thumbnail-sized, but crusts and masses can be palm-sized

COLOR: Green, greenish yellow to yellow, brown

OCCURRENCE: Uncommon

NOTES: In the western half of Colorado lies a portion of the Colorado Plateau, an enormous high-elevation expanse that covers much of the southwestern US. This is the primary Colorado locality for volborthite, a green-to-yellow mineral consisting primarily of copper and vanadium. Volborthite's crystals are very small and generally cannot be seen in most specimens. Instead, it forms as tiny green masses filling the spaces between sand grains in sandstone. Most of the time, identification is easy because of this telltale mode of occurrence, but when a specimen is more yellow in color, it can be confused with several other minerals. Carnotite and metatyuyamunite both occur in Colorado's western plateaus and also form as tiny grains embedded in sandstone. Those two minerals are radioactive, so a simple check with a Geiger counter should distinguish them from non-radioactive volborthite. The only catch is that carnotite and volborthite often occur together in the same specimen, and with no grains large enough to test for streak or hardness, you'll have to make an educated guess when coming up with a label. If you're lucky, it is possible to find very small, but visible, volborthite crystals. They appear as shiny, thin plates or flakes.

WHERE TO LOOK: Western Colorado's plateau region, especially the Uravan Belt in Montrose, San Miguel and Mesa counties.

Hübnerite crystals (dark red)

Embedded hübnerite

Embedded rectangular ferberite crystals (black)

Hübnerite crystals (dark red)

Wolframite series

HARDNESS: 4–4.5 **STREAK:** Brown to black

ENVIRONMENT: Mountains, mine dumps

Occurrence

WHAT TO LOOK FOR: Dark, translucent, rectangular bladed crystals with striated (grooved) faces

SIZE: Wolframite crystals can grow to several inches long, but are generally thumbnail-sized and smaller

COLOR: Black to brown, dark red to red, yellowish brown

OCCURRENCE: Uncommon

NOTES: There are three wolframite series minerals present in Colorado: hübnerite, ferberite and wolframite itself. Hübnerite consists of manganese, tungsten and oxygen, while ferberite is made of iron, tungsten and oxygen. Since manganese and iron can freely exchange for each other in a mineral's composition, hübnerite and ferberite can "mix" to form wolframite, a mineral containing both manganese and iron, along with the tungsten and oxygen present in all three minerals. When this kind of relationship between minerals takes place, it is known as a mineral series. Hübnerite is the most widely collected wolframite mineral in Colorado, and is desirable for its beautiful, deep red, rectangular crystal blades. The thin, slender crystals are often embedded in rock, appearing as randomly oriented reddish black streaks, but finer specimens exhibit free-standing crystals, often alongside brown crystals of scheelite or green fluorite. Ferberite and wolframite are less common and harder to identify because their black, rectangular crystals are often much smaller and less conspicuous than those of hübnerite. In addition, they both tend to form unremarkable black crusts on rock. Hübnerite's crystals are so distinctive that it's easy to identify once you've found it.

WHERE TO LOOK: Hübnerite can be found in Ouray and Gilpin counties embedded in ore veins.

Stilbite

Thomsonite

Chabazite

Loose analcime crystal

Various zeolites on basalt

Heulandite (brown)

Mordenite (white)

Celadonite mica (green)

Zeolite group

HARDNESS: 3.5–5.5 **STREAK:** Colorless to white

ENVIRONMENT: Mountains, plateaus

WHAT TO LOOK FOR: Delicate, light-colored minerals with a fibrous or ball-like structure found forming within pockets in rock

SIZE: Zeolites remain small and crystal aggregates are generally palm-sized or smaller

COLOR: Colorless to white, gray, pale pink to orange, brown

OCCURRENCE: Uncommon

NOTES: The zeolites are a diverse group of complex aluminum- and water-bearing minerals that form in cavities or vesicles (gas bubbles) in volcanic rocks, almost exclusively basalt. There are dozens of zeolite minerals, 25 of which are present in Colorado, though only a few are very prominent. Analcime commonly forms glassy, angular, ball-like crystals, whereas thomsonite appears as rounded, botryoidal (grape-like) masses lining cavities. Chabazite forms small, blocky, square crystals. Stilbite forms odd, bowtie-shaped crystal aggregates. All of these are fairly abundant. Heulandite appears as small, elongated blocky crystals along with natrolite and mesolite, which both form small, slender, needle-like crystals. These can be found in Colorado as well, but they are not as abundant. All of the zeolites are typically light-colored, soft, and frequently occur as well-formed and easily identifiable crystals. As mentioned above, they primarily form within cavities in basalt, so be mindful of any small crystals that you notice when in basalt-rich areas. Some zeolites, especially chabazite, can greatly resemble calcite or quartz, though calcite is softer than all zeolites while quartz is harder.

WHERE TO LOOK: The Table Mountains in Jefferson County are the best known locality in the state for zeolites.

Zippeite (yellow) on coal

Coffinite (black)

Zippeite (yellow)

☢ Zippeite

HARDNESS: 2 **STREAK:** White to yellow

ENVIRONMENT: Mountains, plateaus, mine dumps

Occurrence

WHAT TO LOOK FOR: Thin coatings of bright yellow radioactive material on the surface of rocks in mine dumps

SIZE: Crusts of zippeite are extremely thin, no more than a millimeter or two thick but can be palm-sized in width

COLOR: Light to dark yellow, orange-yellow

OCCURRENCE: Uncommon

NOTES: Zippeite is one of Colorado's many radioactive uranium-bearing minerals. Like several other uranium minerals, zippeite develops a bright yellow color, making identification difficult when compared with carnotite, metatyuyamunite and meta-autunite, all of which are also radioactive. Zippeite forms when uraninite and other uranium-rich minerals weather and decay, lending their components to the creation of new minerals. Zippeite's crystals are always tiny and nearly impossible to see; instead, it often forms as soft, powdery "blobs" on the surface of other minerals, which is one of the few clues to its identity that an amateur can easily observe. Of course, there are also those specimens of zippeite that form as massive crusts atop rock that are virtually indistinguishable from the aforementioned similar minerals. However, there is one test that may yield helpful results: under short-wave ultraviolet light, zippeite glows a bright yellow color, while meta-autunite glows a bright green. Metatyuyamunite and carnotite are not fluorescent. Zippeite's vivid yellow color led to its use as a paint pigment in the mid-1800s. Needless to say, the dangers of radioactivity were not fully understood at the time.

WHERE TO LOOK: Mine dumps in the Uravan Mineral Belt, along the Utah border in Mesa, Montrose and San Miguel counties, are your best bet.

Zircon crystal cluster

Cyrtolite crystal fragment

Zircon crystal (5mm)

Crudely formed zircons (reddish) in quartz

Zircon

HARDNESS: 7.5 **STREAK:** Colorless

ENVIRONMENT: Mountains

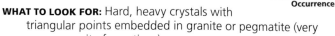

Occurrence

WHAT TO LOOK FOR: Hard, heavy crystals with triangular points embedded in granite or pegmatite (very coarse granite formations)

SIZE: Common grains of zircon are smaller than a pea, but rare, fine crystals can be as large as a golf ball

COLOR: Brown to reddish brown, dark red to purple, gray

OCCURRENCE: Uncommon

NOTES: Zircon is the most abundant zirconium-bearing mineral worldwide and is a very popular collectible in Colorado. Technically, it is a very common mineral, appearing as tiny grains in virtually every kind of volcanic or metamorphic rock. However, specimens of a larger, more collectible size are considerably less common, especially when it comes to crystals. Zircon generally forms as elongated prisms, but in Colorado, elongation of crystals isn't often observed and instead the crystals are more frequently found as small dipyramids (two pyramid-shaped points placed base-to-base). Often several dipyramids are intergrown and form attractive crystal clusters. These zircon crystals most often form embedded in granite or pegmatite (very coarse granite formations), especially in the area around Pikes Peak and St. Peters Dome, in El Paso County. Zircon's color ranges from dull gray to bright, lustrous reddish brown or purple. Identification is fairly easy if the crystal structure is present, otherwise its hardness and occurrence within granite are key characteristics. Cyrtolite, an often crudely formed, brown variety of zircon found in Colorado, contains many uranium and thorium impurities, which makes specimens radioactive.

WHERE TO LOOK: Look for Zircons in the Pikes Peak area, especially in Douglas, El Paso, Park and Teller counties.

Embedded zunyite crystals (brown)

Embedded zunyite crystals (brown)

Triangular crystal face

Zunyite

HARDNESS: 7 **STREAK:** White

ENVIRONMENT: Mountains, mine dumps

Occurrence

WHAT TO LOOK FOR: Pale brown crystals with triangular faces embedded within light-colored minerals

SIZE: Zunyite crystals are rarely larger than a few millimeters in size

COLOR: Colorless to white, gray, light brown to dark brown

OCCURRENCE: Rare

NOTES: Zunyite is one of the many Colorado-type minerals, which means that the first specimen ever discovered came from Colorado—namely the Zuni Mine, in San Juan County, for which the mineral was named. Zunyite is a rare mineral that primarily forms when rocks containing feldspars weather and are altered. The feldspars give zunyite its aluminum content. This process causes specimens of zunyite to become embedded within other minerals, particularly kaolinite and pyrophyllite, which are also produced due to weathering rocks. Zunyite's crystals are glassy, brown tetrahedra (crystals with four triangular faces) that resemble a three-sided pyramid. Often the tips of zunyite crystals appear truncated, ending with flattened faces instead of sharp little points.

There aren't many minerals you could easily confuse with zunyite. Its crystals are a very distinctive shape, and combined with their small size, glassy luster, embedded habit, and brownish to gray coloration, identification is easy. Because of its high hardness, you may initially think it is quartz, but quartz forms in six-sided crystals, not at all like the triangular faces of zunyite.

WHERE TO LOOK: Look in the San Juan Mountains, in southwestern Colorado, where volcanic rocks are plentiful.

Glossary

ACCESSORY MINERAL: A mineral not important to the identification of a rock, e.g. minerals that formed later within cavities in the rock

AGGREGATE: An accumulation or mass of crystals

ALTERATION: Chemical changes within a rock or mineral due to the addition of mineral solutions

AMPHIBOLE: A large group of important rock-forming minerals, commonly with a fibrous appearance

AMYGDULE: A vesicle, or gas bubble, that has been filled with a secondary mineral

ASSOCIATED: Minerals that often occur together due to similar chemical traits

ASBESTOS: A very fibrous, flexible, silky-feeling mineral formation; this can refer to several different minerals, especially chrysotile serpentine and tremolite

BAND: An easily identified layer within a rock or mineral

BED: A large, flat mass of rock that is generally sedimentary

BITUMEN: A black, extremely thick tar-like fluid made of a mixture of liquids formed by organic matter. Can be distilled to produce petroleum

BOTRYOIDAL: Crusts of a mineral formed in rounded masses, resembling a bunch of grapes

BRECCIA: A coarse-grained rock composed of broken, angular rock fragments that have been solidified together

CHALCEDONY: A massive, microcrystalline variety of quartz with a waxy luster

CLAY: Extremely fine-grained sediment consisting primarily of clay minerals

CLAY MINERALS: A loosely defined group of aluminum-rich minerals that occur as microscopic crystals, including montmorillonite, kaolinite, illite and smectite

CLEAVAGE: The property of a mineral to break along the planes of its crystal structure, which reflects its internal crystal shape

COMPACT: Dense, tightly formed rocks or minerals

CONCENTRIC: Circular, ringed banding, resembling a bull's-eye pattern, with larger rings encompassing smaller rings

CONCRETION: A hard, generally rounded shape resulting from sediment accumulating around a central nucleus, particularly fossil material

CONCHOIDAL FRACTURE: The property of a mineral to produce rounded, half-moon-shaped cracks when struck

CONTACT METAMORPHISM: Rock metamorphosis taking place as a result of magma being forced into a body of pre-existing rock

CRYPTOCRYSTALLINE: Crystal structure too small to be seen even by conventional microscope

CRYSTAL: A solid body with a repeating atomic structure formed by the hardening of an element or chemical compound

CUBIC: A box-like structure with six equal sides

DEHYDRATE: To lose water contained within

DENDRITE: A mineral structure resembling the formation of a tree and its branches

DRUSE: A coating of small crystals on the surface of another rock or mineral; also spelled "druze"

DRUSY: Referring to a formation of druse, e.g. "drusy quartz"; also spelled "druzy"

DULL: A mineral that is poorly reflective

EARTHY: Resembling soil, a mineral with dull luster and rough texture

EFFERVESCE: When a mineral placed in an acid gives off bubbles caused by the mineral dissolving

EFFLORESCENCE: A mineral powder or coating produced on the surface of rocks via the evaporation of mineral-bearing water

FELDSPAR: An extremely common and diverse group of light-colored minerals that are most prevalent within rocks and make up the majority of the earth's crust

FIBROUS: Fine, rod-like crystals that resemble cloth fibers

FLOAT: Metal that has been scraped up, moved and rounded by glaciers, resulting in a "nugget"

FLUORESCENCE: The property of a mineral to give off visible light when exposed to ultraviolet light radiation

FOLIUM: A thin, leaf-like structure; the plural is folium

FRACTURE: The way a mineral breaks or cracks when struck, often referred to in terms of shape or texture

GEIGER COUNTER: An instrument used to detect and measure radioactive energy

GEODE: A round, hollow rock formation often containing a drusy coating of crystals within, especially quartz

GLASSY: A mineral with a reflectivity similar to window glass, also known as "vitreous luster"

GNEISS: A rock that has been metamorphosed so that some of its minerals are aligned in parallel bands

GRANITIC: Pertaining to granite or granite-like rocks

GRANULAR: A texture or appearance of rocks or minerals that consist of grains or particles

GROUP: Minerals with similar chemical compounds and crystal structures

HEXAGONAL: A six-sided structure

HOST: A rock or mineral on or in which other rocks and minerals occur

HYDROUS: Containing water

IGNEOUS ROCK: Rock resulting from the cooling and solidification of molten rock material, such as magma or lava

IMPURITY: A foreign mineral within a host mineral that often changes properties of the host, particularly color

INCLUSION: A mineral that is encased or impressed within a host mineral

IRIDESCENCE: When a mineral exhibits a rainbow-like play of color

LAMELLAR: Said of minerals composed of thin parallel crystals arranged into book- or gill-like aggregates

LAVA: Molten rock that has reached the earth's surface

LIME: Calcium oxide, a chemical compound containing calcium and oxygen; generally describes calcium-rich compounds or minerals

LODESTONE: A naturally charged variety of the mineral magnetite; it acts as a natural magnet and will attract iron

LUSTER: The way in which a mineral reflects light off of its surface, described by its intensity

MAGMA: Molten rock that remains deep within the earth

MAGNETISM: The ability of an iron-rich rock to attract a magnet

MASSIVE: Minerals that don't occur in individual crystals but rather as a solid, compact concentration; rocks can also be described as massive; in geology, "massive" is rarely used in reference to size

MATRIX: The rock in which a mineral forms

METAMORPHIC ROCK: Rock that formed after existing igneous or sedimentary rocks were altered due to heat and pressure

METAMORPHOSED: A rock or mineral that has already undergone a metamorphosis

MICA: A large group of minerals that occur as thin flakes arranged into layered aggregates resembling a book

MICACEOUS: Mica-like in nature; said of a mineral aggregate consisting of thin sheets

MICROCRYSTALLINE: Crystal structure too small to see with the naked eye, requiring the use of a microscope

MILLIREM: 1/1,000 of a rem, abbreviated to "mrem" or "mR"

MINERAL: A naturally occurring chemical compound or native element that solidifies with a definite internal crystal structure

NATIVE ELEMENT: An element found naturally uncombined with any other elements, e.g. copper

NODULE: A rounded mass consisting of a mineral, generally formed within a vesicle

OCTAGONAL: An eight-sided structure

OCTAHEDRON: A structure with eight-faces, resembling two pyramids placed base-to-base

ONYX: A mineral mass featuring parallel bands of a mineral, particularly chalcedony (quartz) or calcite

OPAQUE: Material that lets no light through

ORE: Rocks or minerals from which metals can be extracted

OXIDATION: The process of a metal or mineral combining with oxygen, which can produce new colors or minerals

PEARLY: A mineral with reflectivity resembling that of a pearl

PEGMATITE: The lowest portion of a granite formation, where the minerals within the magma are allowed great amounts of time to cool and therefore fully crystallize, often resulting in very large, and sometimes rare, crystals

PHENOCRYST: A crystal embedded within igneous rock that solidified before the rest of the surrounding rock, thus retaining its true crystal shape

PHOSPHORESCENCE: Said of a fluorescent mineral that will continue to emit light from within itself after the radiation source has been removed

PLACER: Deposit of sand containing dense, heavy mineral grains at the bottom of a river or lake

POLYMORPH: Chemical compounds that crystallize in different forms depending on the environmental conditions during formation; for instance, titanium dioxide can form three distinct minerals: rutile, brookite or anatase

PORPHYRY: An igneous rock containing many phenocrysts

PRIMARY: An original formation of rock

PRISMATIC: Crystals with length greater than their width and at least four sides of similar length and width

PSEUDOMORPH: When one mineral replaces another but retains the outward appearance of the initial mineral

PYRAMIDAL: Crystals resembling a pyramid with four or more total faces

PYROXENE: A group of dark, rock-building minerals that make up many dark-colored rocks like basalt or gabbro

RADIATING: Crystal aggregates growing outward from a central point, resembling the shape of a paper fan

RARE EARTH ELEMENTS: A group of seventeen rare metallic elements, named for their low abundance and discovery within uncommon minerals. Includes yttrium, cerium, lanthanum and ytterbium

REE: Abbreviation for "rare earth element"

REE MINERAL: A mineral containing one or more of the rare earth elements. REE minerals can contain variable amounts of several rare earth elements; the element that is most common is noted in parentheses after the mineral name, e.g. samarskite-(Y) is the yttrium-dominant form of samarskite

REGIONAL METAMORPHISM: Metamorphosis occurring in a large amount of rock as a result of great heat and pressure over a large area

REM: A unit of measurement used for calculating radiation's effect on human tissue, most often measured in terms of "rems per hour"; short for "roentgen equivalent man"

RHOMBOHEDRON: A four-sided shape resembling a tilted or leaning cube

ROCK: A massive aggregate of mineral grains

ROCK-FORMING: Refers to a mineral important in rock creation

SCHIST: A rock that has been metamorphosed so that most of its minerals have been concentrated and arranged into thin parallel layers

SECONDARY: A rock or mineral that formed later than the rock surrounding it

SEDIMENT: Fine particles of rocks or minerals deposited by water or wind, e.g. sand

SEDIMENTARY ROCK: Rock derived from sediment being cemented together

SERIES: A small group of minerals with nearly identical chemical compounds wherein one element can freely interchange with another and retain the same molecular structure

SERPENTINE: A group of iron- and magnesium-rich minerals that are generally green in color with a greasy texture

SILICA: Silicon dioxide, more commonly known as quartz

SMELTING: Processing a rock or mineral, usually by melting, in order to separate metals

SOAPSTONE: A metamorphic rock consisting primarily of talc, an extremely soft mineral that feels slippery and resembles soap

SODA: Sodium carbonate, a chemical compound containing sodium and carbon; often used to describe sodium-rich compounds or minerals

SPECIFIC GRAVITY: The ratio of the density of a given solid or liquid to the density of water when the same amount of each is used, e.g. the specific gravity of galena is approximately 7.5, meaning that a sample of galena is about 7.5 times heavier than the same amount of water

SPECIMEN: A sample of a rock or mineral

STALACTITIC: Resembling a stalactite, which is a cone-shaped mineral deposit; carrot-shaped

STRIATIONS: Parallel grooves in the surface of a mineral

TABULAR: A crystal structure in which one dimension is notably shorter than the others, resulting in flat, plate-like shapes

TARNISH: A thin coating on the surface of a metal, often differently colored from the metal itself (see *oxidation*)

TETRAHEDRA: Plural form of *tetrahedron*

TETRAHEDRON: A geometric shape consisting of four three-sided faces; a four-sided pyramid with triangular faces

TRANSLUCENT: A material that lets some light through

TRANSPARENT: A material that lets enough light through as to be able to see what lies on the other side

TWIN: An intergrowth of two or more crystals

TYPE LOCALITY: The location where a particular mineral was first discovered and described

TYPE-LOCALITY SPECIMEN: A mineral specimen collected from the site where that mineral was originally discovered, e.g. the mineral zunyite was first discovered in the Zuni Mine in San Juan County, Colorado; therefore, any sample of zunyite from that location is a type-locality specimen

VEIN: A mineral, particularly a metal, that has filled a crack or similar opening in a host rock or mineral

VESICLE: A rounded cavity created in an igneous rock by a gas bubble trapped when the rock solidified

VESICULAR: Containing many vesicles; a rock containing vesicles is said to be vesicular

VITREOUS: A mineral with reflectivity resembling that of glass

VUG: A small irregular cavity within a rock or mineral that can become lined with different mineral crystals.

WAXY: A mineral with a reflectivity resembling that of wax, such as a candle

ZEOLITE: A complex group of minerals that contain aluminum, silica and excess water; their water content can be lost and regained easily

Colorado Rock Shops and Museums

ACKLEY'S ROCKS
3230 N. Stone Ave.
Colorado Springs, CO 80907
(719) 633-1153

BLUE STONE (rock shop)
4855 W. Eisenhower #C
Loveland, CO 80537
(970) 278-4015

COLUMBINE ROCK SHOP
633 Main
Ouray, CO 81427
(970) 325-4345

DESERT GEMS
457 Wadsworth Blvd.
Lakewood, CO 80226
(303) 426-4411

DENVER MUSEUM OF NATURAL SCIENCE
2001 Colorado Boulevard
Denver, CO 80205
(303) 370-6000

DINOSAUR NATIONAL MONUMENT
4545 Highway 40
Dinosaur, CO 81610
(970) 374-3000

EGGER'S LAPIDARY (rock shop)
16950 S. Golden Rd.
Golden, CO 80401
(303) 279-3952

FLORISSANT FOSSIL BEDS NATIONAL MONUMENT
15807 Teller County Road 1
Florissant, CO 80816
(719) 748-3253

GEORGETOWN ROCK SHOP
501 6th Street
Georgetown, CO 80444
(303) 569-2750

GYPSUM ROSE MINERALS AND FOSSILS
1800 Miner Street
Idaho Springs, CO 80452
(303) 567-2219

MINERAL ADIT ROCK SHOP
2824 West Colorado Ave.
Colorado Springs, CO 80904
(719) 475-1557 (Call ahead)

NATIONAL MINING HALL OF FAME AND MUSEUM
120 West Ninth St.
Leadville, CO 80461
(719) 486-1229

RED AND GREEN MINERALS
7595 West Florida Ave.
Lakewood, CO 80232
(303) 985-5559

RED ROSE ROCK SHOP
490 Moraine Ave.
Estes Park, CO 80517
(970) 586-4180

THE ROCK DOC AT PROSPECTOR'S VILLAGE (rock shop)
17897 US Hwy 285
Nathrop, CO 81236
(719) 539-2019

THE ROCK HUT (rock shop)
706 Harrison Avenue
Leadville, CO 80461
(719) 486-2313

ROCK OF AGES ROCK SHOP
5 Del Wood Dr.
Bailey, CO 80421
(303) 816-7222

Bibliography and Recommended Reading

Books about Colorado Rocks and Minerals

Chronic, Halka, and Williams, Felicie, *Roadside Geology of Colorado*. Missoula: Mountain Press Publishing, 2002.

Eckel, Edwin B., *Mineralogy of Colorado*. Golden: Fulcrum Publishing, 1997.

Mitchell, James R., *Gem Trails of Colorado*. Baldwin Park: Gem Guides Book Co., 1997.

Voynick, Stephen M., *Colorado Rockhounding*. Missoula: Mountain Press Publishing Company, 1994.

General Reading

Bates, Robert L., editor, *Dictionary of Geological Terms, 3rd Edition*. New York: Anchor Books, 1984.

Bonewitz, Ronald Louis, *Smithsonian Rock and Gem*. New York: DK Publishing, 2005.

Chesteman, Charles W., *The Audubon Society Field Guide to North American Rocks and Minerals*. New York: Knopf, 1979.

Johnsen, Ole, *Minerals of the World*. New Jersey: Princeton University Press, 2004.

Pellant, Chris, *Rocks and Minerals*. New York: Dorling Kindersley Publishing, 2002.

Pough, Frederick H., *Rocks and Minerals*. Boston: Houghton Mifflin, 1988.

Index

About the Authors

Dan R. Lynch has a degree in graphic design with emphasis on photography from the University of Minnesota Duluth. But before his love of the arts came a passion for rocks and minerals, developed during his lifetime growing up in his parents' rock shop. Combining the two aspects of his life seemed a natural choice and he enjoys researching, writing about and photographing minerals. Working with his father, Bob Lynch, a respected veteran of the rock-collecting community, Dan spearheads their series of rock and mineral field guides— definitive guidebooks useful for rock hounds of any skill level. Dan's meticulous research allows him to create a relatable text that helps amateurs "decode" the sciences of geology and mineralogy. He also takes special care to ensure that his photographs complement the text and represent each rock or mineral exactly as you'll find them. Encouraged by his wife, Julie, he works as a writer and photographer.

Bob Lynch is a lapidary and jeweler living and working in Two Harbors, Minnesota. He has been working with rocks and minerals since 1973, when he desired more variation in gemstones for his work with jewelry. When he moved from Douglas, Arizona, to Two Harbors in 1982, his eyes were opened to Lake Superior's entirely new world of minerals. In 1992, Bob and his wife Nancy, another jeweler, acquired Agate City Rock Shop, a family business founded by Nancy's grandfather, Art Rafn, in 1962. Since the shop's revitalization, Bob has made a name for himself as a highly acclaimed agate polisher and as an expert resource for curious collectors, helping them to learn about and identify their finds. Now, the two jewelers keep Agate City Rocks and Gifts open year-round and are the leading source for Lake Superior agates, with more on display and for sale than in any other shop in the country.

Notes

Notes

Notes

Notes

Notes